CUBAN SUGAR MILLS

This alphabetical list comprises the 31 sugar mills illustrated in this work. However, as of November 1997 there are about 20 other mills where steam locomotives exist although most of these are out of use, derelict or dumped. In Volume Two, it is hoped to illustrate many of the preserved locomotives scattered throughout the country as they give a fascinating glimpse into the early operations on Cuban railways.

Mill	No.	Gauge	Provine	Page
Mistad con los Pueblos	201	Standard	La Habana	5
Augusto César Sandino	105	3'0"	La Habana	3
Australia	303	Standard	Matanzas	12, 13
Boris Luis Santa Coloma	212	Standard	La Habana	10, 11
Carlos Baliño	446	Standard	Villa Clara	40
Carlos Caraballo	437	Standard	Villa Clara	33
Carlos Manuel de Céspedes	505	Standard	Camagüey	46, 47
Ciro Redondo	515	Standard	Ciego de Avila	48, Front Cover
Cuba Libre	306	Standard	Matanzas	16
Ecuador	504	Standard	Ciego de Avila	44, 45
Espartaco	413	2'6"	Cienfuegos	26, 27
Fructuoso Rodrìgez	312	Standard	Matanzas	17
Granma	304	Standard	Matanzas	14, 15
Ifrain Alfonso	440	Standard	Villa Clara	34 - 38
Jose Smith Comas	315	Standard	Matanzas	18, 19
Juan Avila	320	Standard	Matanzas	22
Julio Reyes Cairo	321	Standard	Matanzas	23
Mal Tiempo	403	2'6"	Cienfuegos	24, 25
Marcelo Salado	428	Standard	Villa Clara	30, 31
Marta Abreu	433	Standard	Cienfuegos	32
Obdulio Morales	418	2'3½"	Sancti Spíritus	28, 29
Orlando Gonzáles Ramirez	503	Standard	Ciego de Avila	42, 43
Osvaldo Sánchez	210	Std/2'6"	La Habana	6, 7
Pablo de la Torriente Brau	107	Standard	Pinar del Rio	4
Pepito Tey	443	2'6"	Cienfuegos	39
Rafael Freyre	635	2'6"	Holguín	Back Cover
René Fraga	319	Standard	Matanzas	21
Rubén Martínez Villena	211	Standard	La Habana	8, 9
Simón Bolivar	448	2'3½"	Sancti Spíritus	41
Venezuela	522	Standard	Ciego de Avila	49
Victoria de Yaguajay	318	Standard	Matanzas	20

All photographs were taken by the author.

INTRODUCTION

It was in the early 1980s that significant numbers of railfans were first able to visit the sugar lines of Cuba more or less unhindered and it was a veritable Aladdin's Cave of ancient steam locomotives that they discovered, their eager eyes happily alighting on sights long thought extinct on the world's railways. There were magnificent 2-8-0s thrashing along out in the fields at the head of 1,000+ ton trains. Lovely little 0-4-0 tank engines switched near the mills and there was a wonderful array of motive power of various gauges in between. Some of the steam fleet was getting on for 100 years old at the time and much of it at least 50 years of age. Miraculously, most of what those pioneering visitors found is still extant in the late '90s and in daily use throughout the season.

This season is inconveniently short for visiting railfans as it lasts only from about Christmas to Easter but for various reasons is totally unpredictable, as is the steam operation in the whole country. Mills do not process cane every day and their trains do not run to regular schedules, so visitors must take potluck and just hope for the best when they call. Diesels have started to appear in recent years and this has slightly reduced the steam level.

Happily, many of the diesels have proved to be somewhat unreliable and diesel fuel is not so readily obtainable as is the low quality, home produced oil burned in the steam locomotives, so there are still systems which are 100% steam, with plenty more having only one or two examples of Rudolf Diesel's invention to make inroads into the steam locomotives' domain. There are some mills which have made a success of diesel operation however and there is even a large, busy mill that has an overhead electric system operating on 1,200v DC but none of these come within the scope of this book. Here, steam reigns supreme!

The Cuban Sugar Ministry, MINAZ, has numbered all the sugar mills from 101 upwards with the lowest numbers in the west of the island and the highest in the east. Similarly, it has put its locomotives into an all-Cuba numbering system. The smallest number, 1101, was also in the west and the highest allocated to the east but since this plan was first introduced there has been a certain amount of intermill movement, so nowadays there are many exceptions. The numbers themselves were also part of a 'Grand Plan'; the lowest series, 11xx, belonging to the least powerful locomotives and the highest group, 19xx, indicating the most powerful engines. Yet again the hand of fate seems to have been at work and there are obvious anomalies here as well but in a way, these just make the Cuban steam scene that bit more fascinating. After all, there surely can be too much regimentation in life.

The selection of photographs which follows is not meant to be a comprehensive overview of steam in Cuba as this would be impossible in this size of book. However, it is intended to be representative, with record shots and shed shots mixed with action photographs and scenic views, all illustrating locomotives at work or at rest on the four different gauges still in use – Standard Gauge 4'8½", 3'0", 2'6" and 2'3½". On the pages which follow, if the gauge is not specifically mentioned, it is Standard Gauge.

With one exception, no mill is featured on more than two pages; the intention being to include as many different examples of the island's steam locomotives in as many interesting and varying locations as possible. Hopefully, this has been achieved but readers must always bear in mind that steam operations in Cuba change constantly and each year visitors find things are somewhat different from the year before.

It is confidently anticipated that steam will remain in service for quite a few years yet but inevitably there will be differences from year to year. It is the discovery of these changes which makes each visit a new experience. The photographs should give the reader at least a flavour of "The Cuban Experience."

David Eatwell

Bedford, England
January 1998

ISBN 0-919487-58-0

ACKNOWLEDGEMENTS

As a mere photographer, it was with some concern that, after agreeing to compile this work, I discovered that it was not just my photographs which were required; lengthy, detailed accurate and informative captions were also expected from me! This presented something of a problem because I am certainly no technical expert where steam locomotives are concerned. But fortunately I know some railfans who are and I have been greatly honoured that so many have been keen to offer assistance in one form or another. Leading the team has been a near neighbour in Bedfordshire, England; George Shuttleworth, who was totally responsible for my being the author, as it was he who suggested to Donald Bain of BRMNA that I could do the job and he who has put me in touch with some additional sources of information that have proved to be absolutely invaluable. I hope that both George and Donald will not think that I have let them down.

Basic details of most Cuban steam locomotives have gradually emerged during the last 15 or 16 years and although I fully believe that what I have written is correct, in some cases there is conflicting evidence. For this reason, I cannot personally guarantee complete accuracy throughout, particularly as continuing investigations reveal new information on a regular basis. I can say, however, that the captions are as accurate as I can make them as they are partly the result of my own observations and also come from careful research through various books, periodicals and technical papers, supplemented by expert assistance from both sides of the Atlantic. If there are mistakes, they come with the highest credentials and the best intentions, but ultimately, of course, I am to blame and I must crave your indulgence if you do find incorrect data. Such items as wheel diameter, cylinder size and tractive effort are mainly taken from manufacturers' lists but it is most likely that these and other dimensions will have undergone a certain amount of unrecorded modification during their years in Cuban service.

Here in England, Chris Capewell, Reg Carter, Graham Leach, Lawrie Marshall, Nick Tindall, Chris Walker, Chris West and Gordon Wiseman have all helped in various ways, as has Allen Copeland in the United States. Publications used for reference include G.A.P. Leach's definitive 1995 book *Industrial Steam Locomotives of Cuba* which lists every known sugar locomotive on the island and is published by the Industrial Railway Society; various editions of David Thornhill's periodical *World Steam*; miscellaneous editions of *Locomotives International* and *Continental Railway Journal* and a selection of specialized papers from different sources.

I am most grateful to all the authors from whose works I have gleaned much detail and to everyone else concerned, including those whose names I may inadvertently have omitted, I proffer my most sincere thanks.

Mill 105, Augusto César Sandino – 3'0" gauge.

The sugar railways of Cuba have never been easy to write about with much certainty for various reasons but mainly due to the constant changes they seem always to be undergoing. Locomotives reported as derelict one year may well be back in service the following year; an all diesel mill this year may revert to steam next year and a mill which was closed one year can be in full working order the next.

Augusto César Sandino offers a perfect example of this because in 1995 it looked derelict with no cane being processed at the mill, rusty rails everywhere and a depressing row of dumped locomotives for all to see by the roadside at the approaches to the mill. Sadly, it seemed that one of the last 3'0" gauge mills in Cuba had closed forever but by the start of the 1996 zafra (harvest), Augusto César Sandino was miraculously back in operation, with much of the track relaid and many of the locomotives overhauled! Earlier, this had seemed to be not just unlikely, but impossible. In Cuba, the impossible happens with the utmost regularity!

This mill is home to the only working narrow gauge Henschel on the island: No. 1405, a 2-6-0 "Mogul" of 1913 plus a total of nine or ten other 2-6-0s and 2-8-0s dating from 1892 to 1923. Eight are Baldwins and, along with the Henschel, there is one Alco. Shown here is No. 1350, a Baldwin 2-8-0 of 1916 returning a loaded train of cane to the mill during the morning of March 29, 1996.

Delivered new to San José Mill as their No. 3, it was still there when the name was changed to Hermanos Ameijeiras after the 1959 revolution and it is thought the locomotive was intended to be used on the Russian-built pleasure line in Lenin Park, situated in the southern outskirts of Havana. This does not seem to have happened, however, and the locomotive came instead to Augusto César Sandino.

With Works No. 43394 and built with 37 in. driving wheels and 15 x 20 in. cylinders, No. 1350 appears to be suffering from quite a severe steam leak from the left hand cylinder area but this will not unduly affect the operations. Without doubt the Cubans will have it stripped down, the gland repacked and have it back in service before the afternoon run. With the barest minimum of facilities and all spares having to be manufactured on site, one cannot help wondering how they do it. But if you want to see miracles performed, spend a little time in the running shed of a Cuban sugar railway. They happen there all the time.

Mill 107, Pablo de la Torriente Brau – Standard Gauge

Only a handful of mills using steam locomotives are situated west of Havana and, of these, the one that should be a *must* on every visitor's itinerary is Pablo de la Torriente Brau.

Two lines serve the mill – a short one to the north and a much longer one to the south – with street running of a sort occurring on both. Like most mills although there are exceptions, locomotives do not turn on this system, so on the northern line, they haul empties out to the loading point boiler first while on the southern line, the more usual method of taking out the empties tender first is employed.

Returning loaded from the south involves a steep climb between the mill-workers' houses accompanied by the inevitable 'thrash' as locomotive fights grade up the last couple of hundred yards or so into the mill yard where a second locomotive is usually waiting to take over the wagons for unloading. This second locomotive may well be one of the gems of Cuba – a Rogers 2-6-0 of 1894, shown (above) by the shed on March 29, 1995.

Built as a tank engine, No. 1501 was originally either a 2-6-2T or a 2-6-4T (records are conflicting) with 48 in. driving wheels and 16 x 24 in. cylinders, and with Works No. 5000 still displayed on the works plates fitted to both sides of the smokebox. So highly do the Cubans value these plates that protective shelves have been installed above them to keep off the rain – at least that is what they look like, although the truth is probably that they are steps to allow crew members and fitters easier access to the boiler top.

When No. 1501 was new, it came to agents Ceballos & Co. for use on the United Railways of Havana and was given the name *CELAYO* but all trace of this is now lost and in common with the other locomotives at Pablo, has its number and the mill name attractively painted on cabside and tender respectively. Letters and figures are in red against a yellow background and are clearly shown in this view of the locomotive raising steam outside its shed on March 29, 1995.

Pablo's pre-revolution name was Orozco, and other locomotives here include a preserved and plinthed Vulcan Ironworks 0-6-0 saddle tank. This is No. 1101 of 1907 – the lowest number in the all-Cuba numbering system. There are also 2-6-0s by Baldwin, Alco, Henschel and Vulcan. Not many mills can boast five different manufacturers amongst their locomotive allocation but Pablo can, and it is no wonder that it has become such a firm favourite with visitors.

Mill 201, Amistad con los Pueblos – Standard Gauge

As with many items of railwayana, the acquisition of locomotive works plates has become an obsession with collectors around the world and even an ordinary plate in average condition can fetch a healthy price at auction. The Burnham, Parry, Williams & Co. Baldwin plate seen here on No. 1106 is one of the rarer examples however, whereas the Extra Order plate of 1923 is one of the not quite so rare, but collectable nevertheless.

A hundred years ago, Burnham, Parry, Williams & Co. ***were*** the Baldwin Locomotive Company, hence the inclusion of their names on the works plate. By early this century, the Parry part of the partnership had been dissolved, and plates dated round about 1900 show the Company names Burnham and Williams. Soon afterwards, even these had been dropped and Baldwin plates from then on had the general appearance and size of the lower plate.

Extra Order plates are not quite as common as other Baldwin plates because they refer to replacement boilers only, and not complete locomotives, hence the difference in the pair of plates fitted to No. 1106. Replacement Baldwin boilers were not restricted to Baldwin locomotives though, and there is a number of locomotives in Cuba so fitted, one of which, No. 1204, is illustrated overleaf on page 6.

Amistad con los Pueblos (literally *Friendship with the Towns* but colloquially *Friendship with the People*) Mill was, until recently, one of a small number of mills where for some inexplicable reason, visitors to the locomotive shed were excluded, and since No. 1106 was hidden away in a dark corner inside, it was not until 1993 when it 'came out' that it could be seen and enjoyed for the first time in many years. The public appearance, although heartily welcomed, was nevertheless tinged with some disappointment, because this attractive little 0-4-0T with the huge cab had only been cosmetically restored and, despite appearances, was not in working order. This restoration had been carried out at a nearby mill and it was here (Osvaldo Sánchez) that it was photographed on April 11, 1993, prior to it going on 'permanent' display at Guines Station, just up the road from the two mills.

Other locomotives which might be found at the now friendly mill at Amistad con los Pueblos are a couple of Alco 2-6-0s of 1905 and 1910 and three Baldwins. They are a 2-6-0 of 1920 and a pair of 2-8-0s of 1919. Two or three of them have usually been found working by recent visitors but those seeking the unusual at Amistad must go to the nearest town to find their 1888-built Baldwin 0-4-0T. When first moved to Guines, it looked nice, but not as nice as it would look if it were operable.

Mill 210, Osvaldo Sánchez – 2'6" and Standard Gauge

Dual gauge running was a common feature on many of the world's railways and even today it is still to be seen on the state systems of such European countries as Germany, France and Portugal, to name but a few, usually combining standard gauge with metre gauge. In Cuba, just a couple of steam mills have mixed gauge although the combination here is standard gauge and 2'6".

One of these mills is Osvaldo Sánchez, situated approximately six miles south of the Great Cuban Highway and about three miles south of the beautiful old colonial-style town of Guines. Examples from the mill are illustrated on this and the page opposite and it is of special interest on at least two other counts: for its modus operandi and for its locomotives. The photographs above and right were taken on April 11, 1993 and there must have been some sort of delay in the operation this Sunday morning because most of the steam allocation was on shed…and in steam! It is standard practice for mills to work as continuously as possible throughout the season. Thus, Sunday is a normal working day, although there are exceptions and it came as something of a surprise to find so many locomotives inactive on this occasion.

Numerous events can conspire to hold up the production of sugar in Cuba. Common causes are derailments of locomotives and/or wagons, mechanical breakdown of the ever-aging mill machinery or locomotives and bad weather. In 1996, the hurricane *Lili* which devastated much of America's eastern seaboard swept across the centre of Cuba and did considerable damage to crops and plant. Other problems include flooding of fields and access roads by Caribbean thunderstorms, loading-point breakdowns, mill washouts (these seem to occur every ten to 14 days) and fuel shortages. It is a wonder that any sugar ever gets harvested here at all!

The shed area was virtually deserted when these photographs were being taken and about the only sign of human activity was the guard on the gate who cheerfully gave his permission for our group to enter the hallowed precincts. Osvaldo Sánchez shed has long been a no-go area and this privilege was swiftly put to good use by the photographers. Among the locomotives present was standard gauge Alco 2-6-0 No. 1507 (seen above) sitting quietly by itself, reassuringly warm and obviously ready and willing to perform any task the absent staff might require of it on their return.

With Works No. 52036, No. 1507 was built at Alco's Schenectady Works in 1912 with 50 in. driving wheels, 18 x 24 in. cylinders and a weight of 109,000 lb. The locomotive went new to Czarnikow, Rionda & Co. as their No. 10 and then came to Providencia Mill where it has stayed ever since. Providencia was the pre-revolution name of Osvaldo Sánchez.

Showing slightly more signs of life in front of No. 1507 is one of the six 2'6" gauge 2-8-0s normally resident here, this being No. 1364 of 1912. Built by Baldwin with the Works No. 37822, No. 1364 had 36 in. driving wheels, 14 x 20 in. cylinders and was imported new by Bernabe Sanchez for use at Senado Mill (re-named Noel Fernández) as No. 9 and named *ALVARADO SANCHEZ*. In common with many Cuban locomotives, some details of the locomotive's later movements are vague but it may well have worked for a while at Sierra de Cubitas Mill before settling down at its present home. If not, it may never become known exactly where it did go, but it has become a regular performer at Osvaldo Sánchez and visitors should have no trouble encountering it.

Of the remaining narrow gauge locomotives here, one is by Alco and the others are Baldwins. The Alco carries the wrong plates which makes identification somewhat difficult but it is thought to have been built either in 1908 or in 1910 at the Cooke Works. Every year new information is discovered, so perhaps it will not be too long before this little bit of missing history comes to light. The four other Baldwins were built between 1912 and 1920 and were mostly serviceable at this time in 1993.

But the main attraction at Osvaldo Sánchez is undoubtedly the very attractive standard gauge 2-4-2T No. 1204, a beautiful little Rogers switcher of 1894 which is shown above. Built with 13 x 22 in. cylinders and 40 in. driving wheels, it was delivered new to the United Railways of Havana as No. 3 , and given the name *PROVIDENCIA*, which probably means it was destined to come to this mill all along! For much of its life, No. 1204 was fuelled by coal but as Cuba has no coal mines, a policy of converting the last few coal-burners (all switchers) to oil was implemented in the early '90s. By 1995 this was almost complete; No. 1204 being converted around 1994.

As mentioned earlier, a replacement Baldwin boiler was fitted in 1919, and the locomotive carries Extra Order plates to this effect, but the oddest feature of its operation is the extremely unusual feat of being able to shunt either standard gauge or narrow gauge wagons around the marshalling-yard as required! The bunker-end bufferbeam is fitted with couplings for both gauges, so it matters not to No. 1204 whether cane is delivered via the standard gauge system or the narrow gauge; it can cope equally easily with both. This photograph was taken on the same Sunday morning as that of No. 1507, and the four crew members were asleep on the footplate with the blinds drawn when the photographer approached but did not seem to mind in the least that their early siesta was being disturbed and willingly threw some coal on the fire for photographic effect. And yes, they were suitably rewarded and, it seemed, were suitably grateful.

Mill 211, Rubén Martínez Villena – Standard Gauge

The best and most interesting collection of small switchers in Cuba is generally agreed to be concentrated at Rubén Martínez Villena, a mill with extensive trackage connected to that of Boris Luis Santa Coloma (*The Mill on the Hill*, see pages 10 and 11) at the Roble flat-crossing, some five or six miles away from the mill buildings.

The pièce de résistance is the 1878-built Baldwin 0-4-2T No. 1112, which is kept in working order, and although it has been steamed once or twice specially for visiting groups, it is probably regarded by the Cubans to be much too venerable to be entrusted with any real work. Still wearing No. 4, its old mill number, on the smokebox door, this early example of Baldwin's expertise as locomotive builders spends most of its days cooped up in an inaccessible shed but a polite request to the right member of staff will usually result in No. 1112 being wheeled out for photography.

Another priceless piece of machinery is what was, at the time the above photograph was taken, one of the last remaining coalburners in Cuba and is No. 1207, a 2-4-0T, of 1912. Shown here pausing between duties outside the fine old Spanish colonial-style mill building on March 31, 1996, this Henschel switcher, Works No. 11677, had been flitting about the mill yard all morning, pushing a few loaded wagons at a time into the mill, and marshalling empties for a larger cousin to take out to a nearby loading point.

Havana agents Schwab & Tillmann imported No. 1207 into Cuba when new, and it came straight to Rosario Mill where it has stayed ever since, and although the mill has now become Rubén Martínez Villena, little else has changed and the locomotive had still not been converted to oil halfway through the 1997 zafra although this minor modification is expected to be carried out with little delay to bring No. 1207 into line with the rest of Cuba's steam locomotive stock.

Henschel locomotive lists are notorious for their paucity of information but No. 1207 is the exception which proves the rule because a few technical details of this locomotive have emerged but of course, being German, they are metric. Weight is 24,500 kg, cylinders are 310 x 430mm and the driving wheels are 900 mm in diameter. Converted to recognizable dimensions and slightly rounded-up, these become 53,900 lb., 12 x 17 in. and 35 in. respectively and we must be grateful indeed for these snippets as little else seems to be known about any of Cuba's other Henschels.

The Vulcan Ironworks of Wilkes-Barre is represented at R. M. Villena by No. 1201, a 0-4-2ST built in 1916 (Works No. 2578), and there is another little Henschel, No. 1206, a

0-4-0T built in 1920. The stock of switchers is completed by Baldwin No. 1311, a 2-6-2T built in 1904 and visitors will be very unlucky indeed not to find at least two of these wonderful little locomotives hard at work during the season. The mill building, a corner of which is shown behind No. 1207, is an architectural masterpiece and makes Rubén Martínez Villena one of the most appealing mills against which to photograph some of Cuba's unique locomotives.

Line work here is in the hands of four Baldwin tender locomotives, three 2-8-0s and a delightful 2-6-0, No. 1411. The 2-8-0s tend to journey to the far reaches of the system, whilst No. 1411 has spent the last few years just working to and from the nearest loading point – a distance of a good quarter of a mile! Built in 1916 (Works No. 44919), it emerged from the Philadelphia factory as an 0-6-0 and came new to USA contractor John Marsch as No. 55. By September 1917 it had been rebuilt into its present form and was then sold to the Cuba Cane Sugar Company as No. 100 for use at the mill which is now known as Seis de Agosto.

Details of the locomotive's transfer to R. M. Villena have been lost but this quite delightful machine has lately been seen in or near the mill, almost invariably coupled to a yellow caboose, even when hauling cane wagons. Adding the front pony truck did not affect other dimensions however and the 16 x 24 in. cylinders still continued to turn the 50 in. driving wheels, although it is hard to imagine the locomotive we see today without its leading pair. It looks 'just right' as it is, so it must have presented a very strange appearance before.

The photograph has all the hallmarks of a real action shot, but in reality the locomotive and train were stationary when the photograph was taken, the smoke having been obligingly supplied for the benefit of the camera, and the breeze doing the rest. The date was March 8, 1995, and the low evening sunshine was perfectly illuminating the motion and wheels. No. 1207 was photographed about midday when the sun was at its highest. This is the reason why some of the under running-plate detail is lost in shadow below the deep side-tank. Unfortunately it is not always possible to do one's photography at the 'best' time of day and one must surely act when one can, or otherwise go without.

Many consider railway photography to be something of an art form, and there is no doubt that in some hands, the camera can be used like a paintbrush, but we are not all O. Winston Link, and most of us must struggle along perpetually against this handicap. However, whilst luck probably pays the major role in any successful steam photograph, luck can be helped along with a bit of planning, and this picture of No. 1411 would probably not have been included here had it not been for the usual kindness demonstrated by the locomotive crew in proving the smoke.

Mill 212, Boris Luis Santa Coloma – Standard Gauge

There are few locations in Cuba where steam-hauled trains cross main roads by over-bridges, it being the norm to accomplish this objective on the flat by the use of level crossings. There are only one or two locations where steam locomotives pass under roads and these are warmly welcomed by photographers as it is rare indeed to find a bit of height from where the camera can be pointed downwards at a train. There is a very attractive bridge on the Espartaco system (page 27) where the sugar line crosses a river but like this overpass, it is equally rare.

Known as the *The Mill on the Hill*, Boris Luis Santa Coloma is just that and was once part of the now electrified Hershey Railway system, yet the reason why such a location was chosen as the site for one of Cuba's larger mills remains something of an enigma. The massive buildings and tall chimneys can be seen from miles around and even if no signposts can be located, first time visitors are able to see from a long way off exactly which way they should be heading.

Sugar mills in Cuba are called *Centrales* (e.g. Central Boris Luis Santa Coloma), and are usually well signposted but in some areas it is not always possible to spot the white chimneys with black tops (resembling huge filter cigarettes), especially when travelling through fields of uncut cane, so these signposts make a welcome sight to newcomers who might otherwise be completely lost and miles from nowhere. Apart from the hard to obtain but excellent 1979 Atlas de Cuba, maps are either completely nonexistent or homemade affairs thoughtfully prepared by previous visitors. Having one is regarded as absolutely essential when looking for a mill for the first time but in the vicinity of Boris one need not worry. Just look up and there it is, completely dominating the surrounding countryside.

Even so, the line itself will not be signposted but if one's approach has been timed fortuitously, another 'aide de navigation' may well be available. At most mills, trains naturally slow down as they get closer but here it is just not possible to do this due to the gradients and locomotives must make their final approach on full regulator, right into the yard, depositing volumes of black smoke into the atmosphere in the process. Away from all mills, steam locomotives can be located by 'smoke-spotting', but nearer to their bases they have a tendency to creep up on the unsuspecting railfan if he is not being vigilant. There is no chance whatsoever of this happening at Boris!

On March 31, 1996, there was a crisis here. No. 1711 had brought up a long train of loads which normally would have been backed into the yard by going all the way over the bridge until the last wagon had cleared the switch and then being propelled into the yard.

Earlier in the day, there had been a serious derailment just out of photograph in front of No. 1711, making such a manoeuvre impossible. In this shot, the locomotive is actually setting back, having already had the train split once and will have to do this three or four more times before all the wagons will be safely deposited in the mill yard.

Derailments in Cuba are not uncommon. They are, perhaps, slightly more frequent on narrow gauge lines than on the standard gauge, but nevertheless the average visitor is likely to encounter at least one or two per visit. Such inconveniences are almost invariably short-lived, however, and this one was true to form. With no shortage of manpower, the derailed locomotive had been rerailed by late afternoon, the badly-damaged track had been repaired and services had returned to normal.

Built in 1920 at the Cooke Works of Alco, No. 1711 (Works No. 62099) was (when photographed) the only 2-8-0 among a stud of five other locomotives (all 2-6-0s) and has been a regular performer on Boris' northern line for some years. With 50 in. drivers (most Alcos in Cuba have 50 in. drivers) and 19 x 26 in. cylinders, No. 1711 weighs in at 135,000 lb., making it well capable of lifting long trains of loaded wagons up steep grades unaided. When it arrived new at the mill which was then called San Antonio, it was their No. 1, but assumed its present number in the all-Cuba numbering system.

The five 2-6-0s are the products of five different factories: Baldwin (1920), the Vulcan Ironworks (1920), and three of Alco's establishments; at Rhode Island (1907), Schenectady (1912) and Cooke (1917) and it is the Baldwin (No. 1604, Works No. 53822) which is shown above forging upwards past the palm trees, only a few hundred yards short of journey's end on the line from the south. Photographed on March 28, 1995, No. 1604 was built with 54 in. driving wheels, 19 x 24 in. cylinders, 170 psi boiler pressure and tractive effort of 27,125 lb. It was delivered new to Gómez Mena Mill where it ran as their No. 6. Gómez Mena was renamed Hector Molina Riano at the big renaming and when it was no longer required there, No. 1604 moved to its present home here at Boris.

Some four or five miles north of the mill is one of Cuba's more unusual features in the shape of the Roble flat crossing (controlled by a tall wooden signal cabin containing a solitary lever and a hand-crank telephone) where three of the four sides have had linking-lines added, turning them into wyes. One wye had been blocked by abandoned wagons for some years but the other two were used regularly in order that trains with their locomotives could be correctly formed and directed for the last part of their journey. In 1997, all three sides were in use. The crossing sees trains from both Boris and Rubén Martínez Villena Mills and will at times be quite busy with trains coming and turning and going in all directions but for real excitement, it is hard to beat locomotives storming up the final approaches to *The Mill on the Hill.*

Mill 303, Australia – Standard Gauge

The story behind the naming of Cuba's sugar mills is almost as interesting as the history of their locomotives and could certainly fill a volume on its own. Most of the original names were swept away after the 1959 revolution and replaced by more 'appropriate' ones in the eyes of the new regime. These include Central Hershey becoming Central Camilo Cienfuegos, San Cristobal being changed to José Martí and Boston being renamed Nicaragua.

The policy seems to have been one of honouring heroes of the Revolution, new politicians and neighbouring friendly states, but perhaps there was an overall shortage of possibilities within these parameters, or maybe purely a lack of imagination, because some mills just had a word or two added to the original title. Cuba became Cuba Libre, Patria was changed to Patria o Muerte, Washington, amazingly, is now George Washington and, equally surprisingly, America became America Libre. One or two enjoyed only slight alterations, but two escaped unchanged – Central Siboney and Central Australia.

With its mill located some 15 miles north of La Bahia de Cochinos (*The Bay of Pigs*) and only a mile or so south of the Great Cuban Highway, the Australia system is one of Cuba's longest, stretching eastwards more than 20 miles and meandering lazily through the flatlands of this part of the country – unkindly described in one guide book as 'Belgium with palmtrees.'

By far the greatest legacy the Russians left behind when they decided to leave Cuba is the Great Cuban Highway which starts at a point towards the western end of the island and eventually peters out in a field miles from nowhere, north of Jatibonico, between Santa Clara and Ciego de Avila. It links up with the Havana ring-road to the south of the city and it was intended when started that it should stretch at least as far as Holguín in the east but it was one of many projects never completed by the former Soviet Union and now the Cubans cannot afford to complete it themselves. But even though it is only a little over half as long as hoped, it does pass fairly close to a number of interesting mills, and this greatly facilitates multi-mill visits by car-borne enthusiasts.

Not that the 'Autopista' is much use to the average Cuban, who is unlikely to have a car, and travelling on such a road can be fraught with danger, especially on those parts furthest from Havana. Besides looking out for some of the biggest potholes on the face of the planet, drivers must also be fully aware that they may encounter pedestrians, pedal cyclists, ox-carts, people on horseback (none of these ever lit at night!), as well as the occasional car, truck or bus, frequently travelling on the wrong carriageway, and which probably will not be carrying any lights either. If nothing else, driving in Cuba could always be described as fun!

At two locations, sugar-lines cross the Great Cuban Highway on the level, one of these being at Ifrain Alfonso Mill (see page 36), the other being here at Australia, and what amazing experiences these are as there never seems to be any concession to road traffic at all while the steam locomotives power their way across at full speed in either direction. A flagman, or flagwoman, should stop what little traffic there is on the road, and 'lives' in an adjacent hut whenever there are any trains on the line...and often when there are not!

Photographers arriving at the crossing, seeking news of the whereabouts of trains on the system are always rewarded with accurate information, often after a quick phone call on the railway's private line and many an unexpected masterpiece has been obtained as a result of this happy cooperation. Trains are naturally few and far between – maybe just one each way a day, sometimes two but sometimes none, so maybe the crossing-keeper is only too pleased to have something to do to relieve the boredom!

The locomotives illustrated here are both Henschel 2-6-0s, and are thought to be the only standard gauge examples from this manufacturer to be regularly involved in heavy line work in Cuba. Most Henschels in Cuba are switchers (two, at Brasil Mill, are fireless), but occasionally other Henschels will make a brief excursion into the outside world and the most likely mills where this might be seen are those at Pablo de la Torriente Brau and Puerto Rico Libre.

On the opposite page, No. 1716 of 1913 was photographed heading towards the furthest loading point with a lengthy train of empties on April 9, 1993 and with the usual extra water car behind the tender. Australia is one of the few mills where locomotives are turned – there is a wye at the mill and another a mile or so from the far end – and this thoughtfulness on the part of the Cubans virtually doubles the photographic opportunities on the system as tender-first pictures are seldom as attractive as those taken when the locomotive is running right-way-round. No. 1716 (whose works plate is inset) came new to the Guedes Sugar Company and eventually reached its present home round about 1989 to join cousin No. 1620 of 1920 (Works No. 18030) and four or five other working locomotives from Alco Schenectady (1) and Baldwin.

No. 1620 (above) is struggling up the grade, probably with more loads than it should be hauling, and assisted by two 'sanders' who have been scooping up dust and dirt off the road to deposit on the track for extra adhesion. This was the second (failed) attempt to reach the summit with this load, after which the train dropped back to the bottom of the grade, deposited a couple of wagons and eventually just managed to struggle over the top at the third attempt. Nothing really unusual about this though, it is a typical occurrence on a typical Cuban sugar railway, but thrilling to the rail fans watching.

Mill 304, Granma – Standard Gauge

One of the many mysteries surrounding the sugar railways of Cuba is how two mills came to be built only a mile or so apart – and not just once! Close to the northern town of Yaguajay is a pair of mills sharing the unusual gauge of 2'3½" – Simón Bolívar and Obdulio Morales (see pages 28, 29 and 41) but 30 miles south of the popular resort of Varadero is another pair: Granma and Victoria de Yaguajay (see page 20).

Like the other couple, these mills are not always processing cane simultaneously but sometimes they are – this apparently fickle behaviour usually depending on the success of the surrounding harvests. Both pairs of mills have linked railway systems and both pairs can operate their locomotives on their partners' tracks. Each mill has its own allocation of locomotives with its own shed and, theoretically, its own route, but it is quite normal to see locomotives from one mill operating on the other's line. If for any reason one of the mills is not working, it is standard practice for the cane to be harvested as usual but then shipped to the adjacent mill (or perhaps one not so adjacent!) for processing, using any suitable and available locomotive.

There is a considerable amount of inter-mill activity in Cuba and with the mills now all state-owned, this is not difficult to organize. Many of these 'irregular' movements will go unreported, of course, because unless a sharp-eyed railfan happens to note something unusual, it will just happen, and no one will be any the wiser. The state railway system (Ferrocarriles de Cuba: F.C.C.) hosts the vast majority of these 'special' workings, some of which take place over considerable distances and not just between closely adjacent mills like Granma and Victoria de Yaguajay.

Although there are no timetables applicable to the sugar lines, there have been certain regular workings over the years. It is always useful to know of these when planning a visit. For example, here at Granma there is a loading point that is only a couple of hundred yards from the mill (see opposite) and for many years a Granma locomotive has been rostered to shuttle over this short distance continuously throughout the day. When a loaded train is ready, it is taken to the mill and, when empties are required, the locomotive delivers them. Nothing seems to imply that this will not continue but this is Cuba and other zafras might see completely different working patterns.

Also, by about 10:00 a.m. daily, a Granma locomotive (but it could be one from Victoria de Yaguajay) came off shed, coupled up to a rake of empties and set off tender first south on the MINAZ track which runs parallel to the F.C.C. in the direction of Jovellanos, dropping off empties at loading points along the way. At Jovellanos, the train did a 'u' turn, headed north

through Carlos Rojas and left its train at a loading point a couple of miles further on. After this, the locomotive returned to Carlos Rojas, set back into the siding and took water while the crew took its lunch, a drink or just a snooze.

About 2:00 p.m., the locomotive went back the few miles to the loading point where loaded wagons were ready for the mill and the return journey started amidst much noise and smoke. Nothing seems to imply that this will not continue either but there have been variations, like those in 1995 when extensive shunting on the Jovellanos wye was necessitated by wagons from the Victoria de Yaguajay Mill, which was not working, having their drop-sides the wrong side for emptying at Granma. In Cuba, it seems, almost anything can happen, and often does!

The shot opposite is Granma's Baldwin 2-8-0 No. 1713 exploding through Carlos Rojas Station on March 9, 1994. The smoke in this instance was not 'by request' – there was no need as these pyrotechnics happen almost every trip. Very few Cuban steam locomotives date from after 1920 but No. 1713 was built as late as 1925. Baldwin built the locomotive for stock and it was sold to Jose Smith Comas when the mill was called Progreso. With 50 in. driving wheels and 20 x 24 in. cylinders, No. 1713 carries Works No. 58486. The locomotive ran at Progreso as No. 4 but has been a popular and regular performer since it arrived at Granma. A total of four 2-8-0s currently comprise the Granma fleet; an Alco of 1916 and one other Baldwin besides the two illustrated, this one of 1920. All have worked regularly up to the present.

Pictured above is another Granma 2-8-0 by Baldwin, shown switching at the loading point nearest to the mill, the stack of which can be seen under the smoke. No. 1812, Works No. 52488, was built in 1919 with 54 in. driving wheels, 22 x 26 in. cylinders, a boiler pressure of 180 psi generating a tractive effort of 35,655 lb. It was ordered by the Cuba Railroad, where it ran as No. 312, possibly doing a stint at España Republicana before settling down at its present home. Locomotives change mills even today and keeping track of them is yet another fascination of the Cuban steam scene.

Of all the mill names in Cuba, Granma probably has more significance to the local population than any other as it was named after the boat (preserved under glass in the centre of Havana) on which Fidel Castro arrived at the island to instigate the Cuban Revolution. The connection with the Communist world is continued by the sight of the Chinese auxiliary generators which are fitted to almost every Cuban steam locomotive; the emblem of Chinese Railways appearing prominently on each end (inset). But auxiliary generators are not the only pieces of Chinese machinery supplied to Cuba. Most Cubans own bicycles and most of their bicycles are Chinese too, sporting such Chinese names as *Forever, Phoenix* and *Flying Pigeon*!

Mill 306, Cuba Libre – Standard Gauge

No. 1808 makes a fine sight as it heads out into the fields with a lengthy train of empties, although getting here has not been all that straightforward. In order to reach this location prior to 1997, the locomotive had to leave the mill tender first at the head of its train, trundle into the F.C.C. station at Navajas where it ran round its consist. It then set off southwards to serve some of Cuba Libre's more distant loading points, negotiating the usual string of unguarded level crossings on the way. By the 1997 season, this complicated manoeuvre had been considerably simplified by the use of a wye.

Vehicle drivers in Cuba encounter a variety of warning signs when approaching level crossings and three different types are illustrated above. In the main picture is the standard crossbuck seen all over the world but in Cuba most usually located along dirt tracks and away from civilization. Inset on the left is the general warning of a mainline railway (F.C.C), while on the right is the sign more usual in built-up areas and along main, metalled roads.

The artists employed to portray steam locomotives on these signs would appear to be in serious need of a familiarization course, which, in a land absolutely bristling with real locomotives, is all the more surprising! But Cuba is far from being alone in this, probably preferring not to 'buck the trend', because from Eastern China to Western Europe and from Northern Russia to Southern Africa, hardly a level crossing sign exists that is anything like realistic and, more often than not, is as ludicrously inaccurate as those shown here. At least there is one good thing about them though – they should be able to make us smile and anything that can do that in this serious world should surely be encouraged.

Five "Moguls" constitute Cuba Libre's steam allocation and it is not at all unusual to arrive at the mill and find every one in steam. The two Vulcan Ironworks examples date from 1916 and 1920 and the other three are Baldwins, two of 1925, and No. 1808 of October 1927. With the Works No. 60212, it is the second youngest Baldwin on the island, while the clear winner in these stakes is No. 1850 at Ifrain Alfonso (page 37).

With 19 x 26 in. cylinders and the usual 50 in. driving wheels, No. 1808 was delivered new to this mill, running as No. 8 until being renumbered, and has been here ever since. The mill buildings themselves have undergone something of a change over the years because here is one of the most massive constructions in all Cuba and being white, it stands out from its surroundings with some distinction. Working inside such a building will not be quite so unpleasant as working inside some of the blackened buildings at other mills, so perhaps Cuba Libre's managers have devised a way to increase output here by keeping the workers happy – or perhaps not!

Mill 312, Fructuoso Rodríguez – Standard Gauge

As noted in the Introduction and as shown at the top of each caption, Cuban sugar mills have been allocated an official number and are being presented here in numerical order, although not quite every mill with steam activity can be included. Despite this, it might be expected that No. 308, the 2'6" gauge mill at Humberto Alvarez should appear next but visitors during the 1993 season were very disappointed to discover that this system with hardworking steam locomotives and particularly pleasant scenery was not in use and the worst was feared. The mill is located only about five miles inland from the popular north coast beach resort of Varadero and one very steep section had often necessitated splitting the train and taking it up in two sections. This doubling of grades is not a common occurrence in Cuba so closure of the system was felt even more.

Subsequent callers have found the mill less and less likely to be reopened but at a recent visit, the locomotive shed was reported still to be housing all the stock and the staff have been kept busy during the intervening period overhauling some of the mill's 2-8-0s; all to no avail apparently, as Humberto Alvarez stays resolutely closed, and seems likely to remain so. But in Cuba you never know – after all, Augusto César Sandino (page 3) has come back from the dead, so why not Humberto Alvarez? All things can be made possible here, of necessity.

The next mill to be featured therefore is Fructuoso Rodríguez and is a mill which has one of the great little locomotives of Cuba on its roster. Line work here might be in the hands of either a "Tenwheeler" or a "Consolidation" built by Baldwin in 1918 and 1919 respectively. Around the yard and the mill, there could be a 1915-built Alco Cooke 2-6-2T working, or the real star of the show, the Rogers 2-4-0 seen above.

Like No. 1501, the Rogers at Pablo de la Torriente Brau (page 4), No. 1216 is of 1894 construction, and another similarity is that it may also have been rebuilt, this time from a 2-4-2T. Some controversy surrounds the works number which has always been thought to be 5001, but which is now believed to be 5036, although there is no argument about driving wheel size or cylinder diameter, these being 40 in. and 13 x 22 in. respectively.

There is no doubt either that the 2-4-0 came originally to the United Railways of Havana when new and that, if the works number is 5036, as now seems probable, then it was unnumbered and was named *SAN AGUSTIN*. But the unusually-shaped tender has given rise to more discussion regarding its origins – one school of thought suggesting that it was constructed like this on site, while another feels that it might have been supplied 'as is' when (or if) the above-mentioned reconstruction took place. Either way, it is believed to be the only one with this shape in Cuba, and greatly adds to the unique appearance of No. 1216.

Mill 315, Jose Smith Comas – Standard Gauge

In Cuba, it is common practice for a locomotive's home mill name to be painted on its tender or side-tanks. The style and decoration is usually left more or less to the whim of the local staff and styles throughout the island vary considerably. As can be seen at Pablo de la Torriente Brau Mill (page 4), some imagination can be used to enhance the final design and nearly always the mill name will appear – but this has not always been the case at Jose Smith Comas. Here, it had long been the name of the nearest town, Cárdenas, which had adorned the locomotives and visitors who were not in the know had every excuse to be confused. In 1997, conformity ruled and the mill name appeared on the locomotives.

But Jose Smith Comas is not the only mill where this deviation from the usual has taken place and 12 miles or so away, another mill in the area has been known to adopt the same habit. Julio Reyes Cairo (page 23) has two German switchers on its books and at least one of them has had the nearest town, Jovellanos, emblazoned on its tank-sides as recently as 1990. By 1993, this had been replaced by a version of the mill name, omitting Cairo, perhaps through shortage of space, or paint! Just to make things more interesting, there is nothing to stop the staff reintroducing 'Jovellanos' whenever they please.

Cuban locomotives are supposed to be painted black, but again poor paint-supplies mean that most of them acquire a dull grey colour fairly soon into the cutting season, even if they were anything like black to start with. Some mills actually keep their locomotives in really smart condition while others do not bother. At Jose Smith Comas they go from one extreme to the other with their switchers like No. 1122 (above) looking absolutely unkempt while the locomotives that do the line work, such as No. 1531 (opposite), are among the best kept locomotives in Cuba.

With the amount of oil leaking from the fuel tank of No. 1122 there is little chance of the motion seizing through lack of lubrication. This leakage was obviously a problem because a few years later the side-tanks had been replaced, possibly as the result of a fire, and the leaks cured in consequence. The locomotive was also anonymous because the replacement tanks were in a dull red undercoat. Although it was now much cleaner, the new colour scheme did nothing to enhance No. 1122's appearance but by the end of the season, the words Jose Smith Comas had been applied.

One of a pair of switchers here – the other is a Vulcan Iron Works 0-4-0ST of 1919 – No. 1122 is an 0-4-0T built by H.K. Porter of Pittsburgh in 1909 with the Works No. 4437.

Agents Kates & Company appear to have been responsible for importing the locomotive into Cuba and by 1948 it had come to Progreso Mill, not being in the least affected when this name was updated to Jose Smith Comas. The diminutive size of the 0-4-0T is clearly illustrated in this side view where it can be compared with the size of the loaded wagon it was taking into the mill when photographed on March 25, 1993. It obviously has more than enough power for the job as the safety valves have lifted, indicating that there is steam to spare, and that cleanliness does not affect performance.

Three other locomotives are regulars at Jose Smith Comas Mill, and Nos. 1415 and 1530 are unusual in that they have had their boilers painted in a smart gingery-brown colour. These are thought to be the only ones in Cuba so honoured. No. 1415, a 2-6-0, came from the Vulcan Iron Works in 1920 but No. 1530 is one of a pair of matching "Moguls" for which Jose Smith Comas Mill is justifiably renowned, the twin locomotive being No. 1531.

This delightful duo came new to Progreso Mill and is its pride and joy. Each locomotive has 18 x 24 in. cylinders and 54 in. driving wheels and with the added, but tasteful, embellishments incorporated on site, might easily qualify as a winner in a 'concours d'élégance des locomotives' competition, should the Cubans ever hold such an event.

Built by Baldwin in September 1925, they have Works Nos. 58654 and 58655 as well as consecutive running numbers and the beautiful condition in which they are kept is shown in the above shot of No. 1531 running out into the fields with a long train of empties on the same day that No. 1122 (opposite) was photographed.

Jose Smith Comas Mill is visited regularly by enthusiasts who like to watch the locomotives being cleaned and polished before every trip. They never venture outside the compound in anything other than ex-works condition and nearly always work 'right way round'. The turning facilities at both ends of the system are normally employed on each trip and it is no wonder that this mill is such a favourite with so many visitors.

While the locomotives are being prepared for the 'regular' daily departure at about 10:00 a.m., the two 2-6-0s make a fine sight sitting side by side in the sunshine on shed and many a railfan has been known to hurry breakfast in order to capture this magnificent occasion on film. By some strange quirk, they are not encouraged inside the chain link fence to do this. However, the ever helpful staff are more than willing to stand back to allow a view through the open gate or to provide a chair on which to stand so that cameras can be raised over the fence! Inexplicable, yet perfectly true!

Mill 318, Victoria de Yaguajay

Joined to Granma Mill by rail (pages 14 and 15), yet thought not to have processed any cane through much of the 1990s, Victoria de Yaguajay Mill has nevertheless kept its locomotives, or at least some of them, in working order to assist nearby when required. The friendly staff are always willing to welcome visitors and show them round the facilities.

As at Cuba Libre (page 16), locomotives haul their trains tender first from the mill, ease forward about a mile to the pleasant little town of Coliseo and pull into the dead end formed when the line that went northwards and connected with Jose Smith Comas Mill (see pages 18 and 19) was closed some years ago. Here, the locomotive runs round, and must be heartily cursed when it does this because the manoeuvre blocks the road and although most of the traffic thus inconvenienced is mill traffic of tractors, cane lorries and occasionally bullock carts, it is not unknown for hold-ups to be extensive.

When it gets going, the train is likely to head off south at a good speed on MINAZ metals that run parallel to the F.C.C. mainline towards the relatively large town of Jovellanos and can then disappear in any of a number of directions, even turning on the Jovellanos wye. A variation to this pattern of working occurred during the 1996 season when problems with the track forced sugar trains to use the mainline for a while. This standby measure was comparatively short lived and things soon returned to normal.

Shown on shed is Baldwin 2-8-0 No. 1813 giving a good impression of being gainfully employed but in fact not doing anything at all; the smoke was being supplied very kindly at the request of the photographer! Built in 1920 with the Works No. 54138, it has 50 in. driving wheels, 20 x 24 in. cylinders and was captured on film in this position at lunchtime on March 24, 1993 before taking out a long train of empties. Coming new to Toledo Mill with the No. 16, it stayed through the renaming to Manuel Martínez Prieto. Then some uncertainty exists because it may have been on the books of Sergio Gonzales Mill before finally reaching Victoria de Yaguajay.

Two other 2-8-0s have seen work from here in recent years, one built in 1920 by the Vulcan Ironworks and the second an Alco that has so far defied all attempts at identification. Only a handful of steam locomotives in Cuba remain undesignated, but research continues. Experts are trying to fill the gaps in the locomotive lists hoping, no doubt, to make a name for themselves with an earth-shattering discovery of momentous proportions.

Mill 319, René Fraga – Standard Gauge

A mixture of locomotives from four different makers, including a mystery Alco as at Victoria de Yaguajay, should be on show at René Fraga. As if that were not enough, one of them is a Henschel switcher and another is No. 1524, a fine example of a medium sized tank engine from the H.K. Porter factory in Pittsburgh, Pennsylvania. Railfans who have travelled in Europe might recognize it as resembling the 2-6-2Ts that worked for so long on the metre gauge Ponferada-Villablino coal line in Spain. Unfortunately, steam has been ousted from this line and it has been diesel-worked for the last few years.

Line locomotives in Cuba are almost always tender engines but once again we have an exception proving a rule as René Fraga's tank engine No. 1524 does line work, besides more mundane duties like switching around the mill. The construction date of this fine looking 2-6-2T was October 1915 and it came new to the J.G. White Engineering Company for work at Mercedes Mill as No. 5, being there at the time of the mill's renaming to Seis de Agosto. In common with most of Porter's products, few builder's details seem to have survived but it is known that No. 1524 was built with 17 x 24 in. cylinders, had a boiler pressure of 180 psi, a tractive effort of 23,065 lb. and that the Works No. is 5741. The photograph was taken when it was dead on shed on March 19, 1995, but it had obviously been working recently.

The Henschel has been out of use for a few seasons now, and is an 0-4-0T of 1913, arriving here via España Republicana Mill but a much more recent arrival is the large Baldwin 2-8-0 No. 1820 which came from George Washington Mill (not illustrated) during 1996, a very late example of an inter-mill locomotive transfer. What makes this so surprising however is the fact that George Washington is approximately 50 miles from René Fraga. Either George Washington really wanted to get rid of No. 1820 or else René Fraga required the locomotive.

There is a cluster of steam operated mills in this area and one of these could have supplied René Fraga with a suitable locomotive if requested. Locomotives from one mill may often be seen working on the lines usually occupied by a neighbour but even more interesting perhaps is the very nicely kept F.C.C. station at Guareiras, through which steam hauled sugar trains pass intermittently. One season they will run through, and perhaps another they will not. Most prominent, and standing guard over the junction 100 yds to the east of the station is a tall, wooden, brightly coloured signal cabin, similar to the one at Roble on the Boris Luis Santa Coloma system (mentioned on page 11) but much flashier and surrounded by a host of semaphore signals. Together they make a very pretty sight.

Mill 320, Juan Avila – Standard Gauge

Not one of Cuba's most extensive systems and with a working method more often found in Java than Cuba, Juan Avila Mill is one of more than a dozen in the Matanzas Province which has a good selection of working steam locomotives to attract enthusiasts. Maybe not the smartest in the country, nor even one with the most interesting engines, Juan Avila nevertheless features a short but very steep climb from the exchange sidings up to the mill which tests the locomotives' pulling powers to the limit and which provides onlookers with almost non-stop entertainment throughout the working day.

Cuba and Java, two of the world's major sugar-producing countries, do not have much else in common, other than their size and shape and their use of steam locomotives sometimes to transport the unprocessed cane back to their mills. Occasionally though, they do have other paths which lead in the same direction and one of these can be found at Cuba's Juan Avila Mill and Java's Soedhono Mill, among others.

It would seem obvious that it must be more economic if the locomotive that hauls the cane from the loading point is the one that takes it right into the mill, rather than use a second locomotive. Both Juan Avila's and Soedhono's masters think otherwise! Why use one when we can use two? It must be their philosophy, because both mills use one locomotive to haul the cane from the loading point and then leave it in an exchange siding a quarter of a mile from the mill. A second steam locomotive comes along later to complete the journey! The reason for this system probably is the use of the exchange sidings as a short term storage point until the mill can cope with cane that is being harvested. Also, the geography of the mill necessitates the sidings being some distance away. At Juan Avila, the locomotive being used for the quick dash up the hill on April 1, 1996 was 2-6-0 No. 1720, sporting one of the largest smokebox number plates in Cuba.

Built by the Vulcan Ironworks in January 1921, No. 1720 bore Works No. 3138. It had 50 in. driving wheels, a boiler pressure of 180 psi, 19 x 24 in. cylinders, a tractive effort of 26,510 lb. and weighed 130,000 lb. With the No. 6, the "Mogul" was originally a Cuba Libre locomotive, arriving there when that mill was named Cuba and surprisingly, must have been surplus to their requirements. Today Cuba Libre can often be seen using its complete complement of five locomotives at any one time during the season. Juan Avila has six other locomotives, among them two Baldwin "Tenwheelers" of 1906, a 2-6-0 twin to No. 1720 from the Vulcan Ironworks and three "Consolidations" (two by Baldwin and a Schenectady Alco). Truly a good selection.

Mill 321, Julio Reyes Cairo – Standard Gauge

Here is one of Cuba's real railway treasures, not quite comparing with Rubén Martínez Villena, perhaps, but a treasure all the same and the pictures above go some of the way to show why.

Cuba still contains a wide variety of locomotive types from a dozen of the world's great locomotive builders and although most of them are American, three are German and one is English – the sole British survivor having been ignominiously converted to a now derelict diesel. Orenstein & Koppel still have one or two examples in use (page 26) and representatives of the other two makes are on display at Julio Reyes Cairo. The first is the only Borsig known still to be in Cuba, an 0-6-0T No. 1123 of 1910 and the second is the exquisite little Henschel shown in the two photographs on this page.

Inset is No. 1124 as it was on March 9, 1994 when still a coal-burner, nicely displayed outside Julio Reyes Cairo shed in beautiful condition, possibly at the behest of a visiting British led group specializing in the smartening-up of 'tatty' locomotives. A year later, as seen in the main photograph, No. 1124's appearance had degenerated alarmingly, it once again having become a working locomotive, as opposed to a show-piece, and was earning its keep marshalling trains as well as doing general switching duties around the mill.

Seen and photographed on March 18, 1995, the 2-4-0T was built at Kassel in 1913 with the Works No. 12445 and delivered new to the now closed La Julia Mill, Melena del Sur. Then, having been put out of a job, it came to Julio Reyes Cairo but in common with the few other coal-burners on the island, did not see a great deal of work due to the inevitable shortage of fuel as Cuba has to import all its coal. Converted to oil for the 1995 season and kept reasonably busy throughout, No. 1124 was not heavily used the following year, being laid up for long periods like its Borsig mill-mate, while diesels did the work previously handled by these two capable little German switchers. Stripped-down and completely overhauled in 1994, the Borsig was a major attraction at Julio Reyes Cairo, providing, with its Henschel partner, a very good reason to visit what would otherwise be described as one of Cuba's more ordinary mills.

Other locomotives here are by Baldwin (2-6-0 and 2-8-0), Alco Cooke (2-6-0) and Vulcan Ironworks (2-8-0), with one of them usually in use as a stationary boiler supplying steam to the mill; their builders' dates span 1915 to 1925. Lying close to the shed is a dumped and derelict Baldwin 2-8-0 of 1891 which, if working, really would be something to shout about as it had been built as a Vauclain compound, and there are not too many of those running on the world's railways in this day and age!

Mill 403, Mal Tiempo – 2'6" gauge

Along with the nearby Espartaco and the somewhat remote Rafael Freyre, Mal Tiempo is the third of Cuba's 'Big Three' narrow gauge systems, but to many observers it is definitely not the third in attractiveness. Nor is it third in convenience, because while it is probably true to say that most rail fans who come to Cuba will visit Mal Tiempo and Espartaco Mills, it is also probably true that most will not venture all the way to the eastern end of the island where, arguably, the only mill of any consequence is Rafael Freyre (see the Back Cover).

Unarguably, though, Mal Tiempo is spectacular, not in the sense of being mountainous, but by being hilly enough to stretch the powerful little Baldwin 2-8-0s to the full and when fully stretched they well and truly let all and sundry know all about it! This is the mill where locomotives cause the ground literally to shake and trains approaching from afar can first be detected not by sight nor sound but by feel! Yes, at Mal Tiempo, the earth really does move!

The usual translation of the mill name is 'Bad Weather' but climatic conditions have nothing to do with movements under foot although the reason behind the name has given rise to some speculation because the weather here seems to be no different from anywhere else in this area of Cuba. In 1996, Hurricane *Lili* ripped much of the roof off the shed but then it also cut a swath right across Cuba and did not single out Mal Tiempo for special treatment. But who can rightly complain about Caribbean weather? For some folk it might get a little hot, for others, sometimes a little breezy and it cannot be denied that there are the occasional thunder-storms some afternoons but bad weather in this part of the world? Not very often.

The reason for the 'shaking' phenomenon lies in the design and construction of the collection of locomotives in use at Mal Tiempo and has nothing at all to do with the terrain or landscape as has been suggested by some 'experts'. Rather, it is caused by the position of the locomotives' burners, all of which are mounted very low, which in itself is strange because of their different histories and the extended period over which they were built. It is thought that they must have been modified here to a 'local design'! Of the half-a-dozen locomotives likely to be usable, only two date from the same year (1920), and the earliest dates from as far back as 1903! All are Baldwins.

As trains are lifted over the many summits on the system, the resultant shock-waves from the hardworking locomotives precede them by miles, causing the earthshaking effect which really has to be experienced to be believed. Users of tripods beware – hand holding your camera at Mal Tiempo will probably result in sharper pictures than using the ground to support your photographic equipment!

Everything about Mal Tiempo is delightful, even to the extent that there is hardly any tender first running. At the mill there is a wye just outside the yard. At the far end of the line

there is another wye adjacent to the loading point and both are used regularly. Locomotives have to work hard at frequent intervals whether they are taking out empties or returning with loads, so there are nearly always dramatic smoke effects to be captured on film and these, combined with the continually undulating track, have provided many a photographer with a mini-masterpiece when combined with the glorious wide open vistas. This is a mill where photographers feel they just cannot lose out.

'Normal' conditions require three locomotives in steam, any or all of which might be out on the line or, as on April 7, 1993, on shed (opposite page). A breakdown somewhere at the mill had caused a temporary cessation in cutting and a consequent cessation of train movements, hence the three 'live' locomotives being in a state of total inactivity that morning. As there was a promise of renewed movements later in the day, arrangements were made to return when there would be more to see but not before the shed personnel had obligingly created some 'atmosphere' for the camera. In Cuba, if you ask, you usually get.

Left to right, the three 2-8-0s are Nos. 1345, 1321 and 1320, dating from 1920, 1913 and 1909 respectively. All were built with 36 in. driving wheels 14 x 20 in. cylinders and are of similar shape, size and appearance although the capped stack of No. 1345 gives this particular machine a more distinguished look than any of its fellows. All have had other homes previously, the centre and right hand locomotives having been at Guillermo Moncada Mill before reaching Mal Tiempo.

Above, No. 1355 (Works No. 53864) is climbing steeply away from the mill's most distant loading point at Potrerillo late in the afternoon of March 22, 1995 with seven loaded cane wagons, an extra water tender and a caboose – about the maximum the 2-8-0 can manage up this heavy incline. The locomotive's vital statistics are the same as the three opposite, as is the roar from underneath when it is being 'thrashed', as here. No. 1355 was, like No. 1345, built in 1920 but a month earlier (in October). Locomotives' relative ages cannot be deduced from their running numbers.

Night-time running is common in Cuba but not regularly employed at all mills. The gleaming glasses of the headlamps on the four locomotives illustrated on these two pages indicate that Mal Tiempo is one mill where work does go on after dark and it will have been noted that this has not been a feature of every locomotive pictured so far – see No. 1720 on page 22, for example. Headlights vary considerably in size and appearance throughout the island, some even being 'twin' lamps (page 14) as commonly appearing on locomotives in Southern Africa and when powered by their 'new' Chinese generators (page 15), can project a beam a considerable distance in front of the train.

About halfway between the large city of Cienfuegos, Cuba's second port and main sugar outlet on the southern coast, and the historic old centre of Santa Clara lies the little town of Cruces. The town's most serious claim to fame might be regarded as the fact that it has a station on the railway line connecting the other two towns. But in the world of railways, Cruces is a town of renown, being near the centre of one of the highest concentrations of steam-operated sugar mills in Cuba. No fewer than six mills are within easy reach and although some of them see steam usage only occasionally, probably when the diesels fail or are being serviced, together these mills form a major part of the steam scene in this part of central Cuba.

Travelling southwest off the Great Cuban Highway on the road to Cienfuegos, the first mill to be encountered is Ifrain Alfonso (pages 34-38), then Carlos Caraballo (page 33) shortly afterwards. A little further along are Ciudad Caracas (not illustrated) to the west and Marta Abreu (page 32) to the east. Mal Tiempo (pages 24 and 25) is on the southern outskirts of Cruces. A further five miles on through open countryside and one of Cuba's 'Big Three' narrow gauge systems is reached at Espartaco.

With lines serving the mill from both the north and the south and with trains liable to operate at almost any time, it is not unusual for three of the mill's 2-8-0s to be in use together, as at Mal Tiempo. They are not required to do their own switching as Espartaco is one of only three or four mills that uses fireless locomotives to perform this task. There are two of them here, both normally in daily use. Shown on the left is No. 1130, an 0-4-0F built by Orenstein & Koppel in 1914 with Works No. 6854. This locomotive has its cylinders in the 'European position' under the cab. As a new unit, this locomotive came here when the mill was named Hormiguero and it has been here ever since. It was rated by the manufacturers at 60 horsepower and was given the running number 0!

Two years later, World War One was in progress, preventing the mill owners from going back to Germany for a repeat order, so they went instead to the USA, purchasing No. 1131 from Baldwin (Works No. 43277) in 1916. Obviously they thought it was not at all strange for a locomotive to be given No. 0 so the new one became No. 00! Being a Baldwin, it looked far more conventional to North American eyes as its cylinders were at the front, just like any other 'normal' engine. It is interesting to be able to compare the different designs in these two pictures, No. 1130 being photographed on April 8, 1993 and No. 1131 (right) on March 12, 1994.

Mill 413, Espartaco – 2'6" gauge

These are both ‘snatched’ shots, taken before the photographer was ejected from the premises by an armed guard but how the old order changeth! In 1996, visitors were welcomed here with open arms and given leave to roam at will, photographing whatever they pleased. In the ’80s when rail fans started exploring the sugar mills on the island, they were turned away from most mills and were welcomed only at a small number; now it is the opposite with not more than half a dozen or so excluding visitors and the number is decreasing every year. At this rate, soon it ought to be none, although there has never been any ‘hassle’ out in the fields and away from industry.

Four 2-8-0s reside at Espartaco, with the ‘star-attraction’ dating from 1895 and the rest having been built between 1911 and 1919 to a design which changed little over the years. Keen eyed observers will notice detail differences with cylinder design and boiler mountings being updated, but the Baldwin 2'6" gauge 2-8-0 with 36 in. driving wheels and 14 x 20 in. cylinders was so successful that there was no need to improve it, so Baldwin did not change it.

The furthest loading point from the mill is at the end of the line which runs a little way past the delightful little town of San Fernando de Camarones. Outside the town is one of those great rarities on Cuba’s sugar lines; a bridge, and on the afternoon of April 8, 1993, there was a loaded train crossing it, hauled by the 1911-built Baldwin 2-8-0 No. 1327, Works No. 37142. Having spent all its life on the system, it must have crossed over this river thousands of times, starting when it carried the number 6, and when the mill still had its original name.

Parts of the Espartaco system are remarkably steep, which limits the loads these powerful little 2-8-0s might otherwise bring back to the mill and at one point on this particular branch there is a gradient so steep that trains come almost to a stand at the summit, despite ‘taking a run at it’ from a good way off. All oil-burners can produce great smoke-effects when required but up the climb at Espartaco they are something extra special and are there without being requested!

The pronunciation of the current name may come as a bit of a surprise to those not familiar with proper nouns in Spanish because when spoken, the emphasis is not on the second syllable (Es-**par**-ter-co) as might be expected, but on the third, making the result Es-per-**tar**-co! Foreign languages are full of such idiosyncrasies though and not the least English – as spoken on both sides of the Atlantic!

Mill 418, Obdulio Morales – 2'3½" gauge

Long thought to have the unique gauge of 2'3¾", it is now realized that whoever first wielded a tape measure across the railway tracks in this part of Cuba, did not read it very accurately because the result announced to, and believed by, the world was wrong and by as much as ¼ in. too! That such a mistake should have been made, and perpetuated, is all the more surprising as it was well known that the complete batch of Baldwin locomotives that was delivered here new was not built to 2'3¾" gauge at all but to 2'3½"! Perhaps the ancient measurer thought the 2'3½" locomotives had all been regauged!

One possible explanation for the discrepancy might be that 2'3¾" approximates to 700 mm, a popular gauge on other sugar railways, notably in Indonesia, but the conversion is not exact, so this may be just speculation. What is well known however, is that the present 2'3½" gauge systems at Obdulio Morales Mill and the adjacent, rail connected, Simón Bolivar Mill (page 41) were part of a much larger public railway which connected the town of Caibarién, some 20 miles west along the coast, with Chambas, another 20 miles to the southeast.

The centre portion of this old railway is now operated by MINAZ, yet there is thought to have been a plan in the early '90s to convert it to standard gauge, although only a few standard gauge rails can be found in the vicinity of the mill today and like many other grandiose schemes in the country, this one also seems to have foundered. Of course, railfans are highly delighted about this because not only do the adjacent mills constitute Cuba's sole 2'3½" gauge railway, the locomotives here would obviously have had nowhere else to work and soon after being laid up, would have become rusty and derelict hulks. Not a happy thought.

Obdulio Morales shed building has been rebuilt on a new site in recent years, now seeming to cater more for the line's diesel fleet than it does for the steam, because while the diesels are being serviced in the modern surroundings, the steam locomotives are more usually sitting outside, almost as if it had been decreed that the old steam locomotives must not be permitted to interfere with the new diesels. Fortunately for steam fans, the diesels seem to require considerably more attention than their elderly relatives and are generally to be seen monopolizing the shed while the steam locomotives have to sit simmering outside. Occasionally though, one is allowed to venture inside but when this happens it is almost as if it knew it should not be there and it will scurry out at the earliest possible moment.

Differing somewhat in appearance from many of the other narrow gauge locomotives in Cuba, all the Obdulio Morales engines working today are Baldwins although one of the half

a dozen dumped and/or derelict machines littered about the place was built by the Vulcan Ironworks. The six locomotives currently working, or workable, were all manufactured between 1919 and 1921 and pictured on these two pages is a pair still wearing their old mill number on their smokebox doors as well as the MINAZ number on their cab sides. No. 1354 (left) shows that it had been No. 7, and the number 2 on locomotive No. 1366 (above) can just be made out.

Photographed on March 23, 1995, about halfway through its journey to the mill from the far loading point with a train of full cane wagons, No. 1354 had the 'standard' Baldwin narrow gauge specifications of 36 in. driving wheels and 14 x 20 in. cylinders when constructed. In view of all the other alterations that have taken place since, it would be a surprise if these have not been 'modified' too! Originally built to 2'7½" gauge with the Works No. 55127, it may well have had a spell as a 2'6" gauge locomotive before coming here and being altered again to 2'3½". Delivered new to Argeo Martinez Mill, when it was called Esperanza, its first move was probably to Diez de Octubre when that mill was operating a dual gauge system, eventually reaching its third (and final?) home at Obdulio Morales.

The not too-steam-tight locomotive shown above is No. 1366 (Works No. 52103) and prior to reaching its present home it had the same history as two of the locomotives illustrated earlier – No. 1364 at Osvaldo Sanchez and No. 1320 at Mal Tiempo. It was No. 12 at Senado Mill which was later renamed Noel Fernández. The photograph was taken from the highest point on the nearest loading point to the mill and shows the MINAZ 'teepee' (wigwam) emblem on the tender instead of the more usual mill identification, a trait shared with the mill next door. Wyes allow most workings at Obdulio Morales to be undertaken with the locomotives running smokebox first.

After No. 1366 had stormed off with its long train of loads and the photographer had set off by car in chase, it took some ten minutes to catch up with the train – but what a forlorn sight it was. Whether the friendly crew had departed with the intention of racing the car-borne photographers back to the mill or not we shall never know but without doubt if it were a race, the Cubans won hands down, but at what cost! The locomotive was sitting sadly at the entrance to the mill yard with a tangled mess of metal which until so recently had been its right-hand motion! The train was blocking the only rail entrance to the mill from this direction, and it was a very crest-fallen crew who had to oversee the dismantling of the broken bits before No. 1366 could be towed out of the way for repair but, typically, they never stopped smiling!

Mill 428, Marcelo Salado – Standard Gauge

The seaside town of Caibarién, although not really a resort, is nevertheless an extremely pleasant place to visit, situated as it is on the north coast some 30 miles due east of Santa Clara. Approximately six miles inland is an even more agreeable place to visit, the beautiful little town of Remedios, with its colonial style buildings set round the old square. As if these attractions were not enough to bring tourists hurrying to the area, halfway between the two towns, right beside the main road is the mill with three saddle tanks. Not only has it three of this type of locomotive but all three have different wheel arrangements. In addition, the mill operates a locomotive which has been described as the best looking 4-6-0 in Cuba.

Built in 1911 with the Works No. 37172, this shot shows the Baldwin "Tenwheeler" No. 1342. This handsome beast with its 56 in. driving wheels, 17 x 24 in. cylinders and sleek taper boiler was caught on camera on March 28, 1993, high-stepping beside the main road just after leaving Remedios with a lunch time train of loaded cane wagons bound for the mill. There are only five or six places in Cuba where steam hauled sugar trains run parallel with asphalted roads and the long stretch between Remedios and Marcelo Salado Mill is probably the best as it offers virtually uninterrupted views of the line over the whole length.

Car-borne photographers are sure of satisfying results and, as the line runs east to west with the road on its south side, the sun will be suitable for photography most of the day. There are no turning facilities on this system, so loaded trains on their way back to the mill will have their locomotives operating boiler first and some of the fastest running by steam hauled trains in the whole of Cuba can be experienced along this level and straight stretch of railway line.

Carlos Baliño Mill (page 40) and the four steam mills in the Ciego de Avila area (pages 42-45 and 48 and 49) probably offer the best alternative opportunities to photograph steam locomotives in motion from a moving car. However, the light can be tricky at these locations and the number of steam hauled trains actually using the roadside stretch at Carlos Baliño has been significantly reduced of recent years. In consequence, for 'pacing' a train by car, it will be hard to find a more rewarding line in Cuba than that which leaves Remedios in the direction of Marcelo Salado and with the generally low level of traffic that exists everywhere on the island, video and movie photographers should come away from here feeling well satisfied.

Not many mills can boast that all their working steam locomotives have different wheel arrangements but Marcelo Salado has this distinction. There is the usual dumped wreck out the back but there are three mainline locomotives for use here and, like the three saddle tanks already mentioned, they are usually kept looking quite smart. The two large locomotives not illustrated are a Baldwin "Mogul" of 1916 and an Alco Cooke 2-8-0 of 1920. It was this machine, looking as pristine as any in Cuba, which, in 1995, was seemingly being used solely to give driving instruction to visiting railfans! The reality is obviously somewhat different but the mill was not working during much of the 1995 season yet some of the allocation was in steam and, in typical Cuban fashion, visitors were even encouraged to climb into the cab of No. 1549 to try their hands at the controls. This was only for short distances up and down the yard but, nevertheless, a 'hands-on' experience never to be forgotten.

Two of the mill's saddle tanks are shown on this page, the missing one being the 1904 Baldwin 2-6-2ST No. 1343 which arrived here comparatively recently, being based previously at Héctor Rodriguez and George Washington Mills. While at the former, it was renumbered from the much less appropriate 1160 and, in view of its comparatively large size, it must be questioned why it was given the lower number in the first place! It is no use trying to find a sensible explanation for much of what goes on in Cuba, it is much better just to accept and enjoy it.

Shown here are 0-4-0ST No. 1147 and 0-6-0ST No. 1148. The former was photographed on March 18, 1994, and was built by Davenport with the Works No. 1738 in 1919, being one of only three known examples by this maker in the country. Its boiler pressure was 170 lbs psi and its tractive effort a mere 10,400 lbs when built. It is the only Davenport to see regular work in Cuba and has been at its present home since at least 1948. A 3'0" gauge example is preserved at Quintin Banderas Mill while the third is at Carlos Manuel de Céspedes Mill (pages 46 and 47) but has been out of use and hidden in a dark corner of the shed for the last few years, seemingly destined not to be allowed to see the light of day ever again. However, it recently has been taken out and it would be nice if it were to be returned to steam but, to date, this does not appear very likely to happen.

No. 1148 was built by Alco Cooke (Works No. 58942) to 750 mm gauge for Russia in 1919 but never reached there and was subsequently regauged. Photographed at work on the same day as No. 1342 (opposite page), it was originally an 0-6-0T and was rebuilt to its present form in 1925, coming to Marcelo Salado as No. 7 under the mill's previous name of Reforma. Its driving wheels are no more than 30½ in. diameter and its cylinders only 11 x 16 in. – its original gauge being the reason for the tiny cylinders. The huge cab gives the locomotive a very front heavy look while the stovepipe stack does little to enhance its overall appearance. Yet it is showered with affection by the staff and, like the other locomotives here, always has a neat and tidy look to greet callers. It is sometimes nice to see locomotives covered in grime to indicate that they are real work horses but it is also encouraging to see some smart ones occasionally. Here, at Marcelo Salado, that is certainly what you should see. But will it last? We can only hope.

Mill 433, Marta Abreu – Standard Gauge

If not the outright winner of the title for The Ugliest Locomotive in Cuba, this fireless engine would certainly be a close runner-up, although which would be the eventual victor might be open to lengthy discussions as there are other fireless locomotives at other mills with very strong claims! But then fireless locomotives are rarely pretty – they are built to contain steam like a mobile storage tank and not to generate steam with a fire and a boiler like a conventional locomotive. The 'boiler' of a fireless locomotive is charged with steam from the factory where it works and of course the length of time between these charges is strictly limited.

About 50% of the boiler space is taken up with steam under pressure and 50% with near-boiling water. The water serves two purposes – it provides some tractive weight to what would otherwise be an engine with little or no adhesion and also it helps to keep the steam as hot as possible for as long as possible in order to lengthen the times between each refill. Many sites where fireless locomotives work require them to return to base to be topped up, but not Marta Abreu. Here, their sole working steam locomotive, No. 1239, need only make its way to the edge of the yard to take on more steam from a pipe coming from the mill. This is a considerable time-saver as well as being more economical with the steam but, despite all this thoughtfulness, nothing has been done to try to make the locomotive look any better. But then it is doubtful if anyone could succeed in this endeavour, no matter what they did!

Con Cuba para todos los tiempos, shouts the slogan on what would be the front of the smokebox on a normal steam locomotive. This literally means *With Cuba for all the times*. Slogans of every kind proliferate in Cuba and this one might be taken to mean *Cuba for ever*, a philosophy shared by most members of this defiant nation. *Yes with Cuba* is another popular slogan but politics do not seem to rate very highly on the priority lists of the average Cuban, so one must wonder how much effect all this 'sloganizing' really has!

In spite of the "smokebox" number plate proclaiming that the locomotive *was* built by Alco in 1919, No. 1239 was constructed by Baldwin in 1921. The 0-4-0F (Works No. 54270) was delivered new to Marta Abreu Mill in pre-revolution days when it was called San Francisco and has worked here ever since. Its boiler pressure is 40 lb. psi and its tractive effort is only 6,800 lb. Not in use every season, nor even all through any season, it does work and is shown as it was on March 13, 1994. It was reported not to have worked at all during this season but to have spent the whole time sheltering from the sun under the corrugated iron roof of its very basic home. In 1996, No. 1239 saw much work, and its place in the shade was taken by a big diesel, probably a failure which could not easily be repaired. Or is that wishful thinking?

Mill 437, Carlos Caraballo – Standard Gauge

Like Marta Abreu Mill (opposite), Carlos Caraballo is another one of those located down 'The Road to the Mills' as described on page 24. Although finding the mill is easy, finding a steam locomotive working here is more difficult. Two derelicts are dumped out the back, a Baldwin 0-6-0ST and an Alco 2-6-2T but there is so little left of them now that it is surprising that they have been identified as accurately as they have. They are thought to be Nos. 127 and 145 respectively and since these are pre-revolution numbers, they give some idea of how long they have been discarded. Even in Cuba it would appear impossible for them ever to be returned to service.

A matching pair of Baldwin "Moguls" is allocated to Carlos Caraballo, yet steam working is scarce and MINAZ diesels are the more usual motive power. But overhauls of steam locomotives still get carried out in the shed and No. 1556 has been seen receiving attention up on blocks during recent visits. Nobody has appeared to be in too much of a hurry to finish what they have started, however, but maybe visitors will soon find that No. 1556 is off its blocks and available for use again which will be a very pleasant surprise.

During the 1993 and 1994 seasons, 2-6-0 No. 1550 was spotted working occasionally out on the line. This is how it looked on April 7, 1993 and surely no locomotive could be turned out in any smarter manner, anywhere. The best way to describe it is immaculate and even when out of use it retained its perfect-looking image until the 1997 zafra when it had a bedraggled appearance. Until then, shedmasters at Jose Smith Comas (pages 18 and 19) and Marcelo Salado (pages 30 and 31) had good reason to eat their hearts out.

When Baldwin built No. 1550 in 1920 with the Works No. 53638, it had 18 x 24 in. cylinders, 50 in. driving wheels, 180 psi, and a 22,032 lb. tractive effort. It has spent all its life here, arriving ex-works when the mill was called Santa Catalina. Appropriately it was No. 1 then but unfortunately there is no record as to why it was given the name *JULIA*. There is no harm in speculating, of course, and it would be nice to think that Julia was the first driver's wife, perhaps, or daughter, or girl-friend!

Prominent on the right and just in front of the pure white smokebox is the mill chimney stack, careful inspection of which will explain an earlier comment that finding a mill is not difficult. It is easy to find any mill in Cuba just by looking around, because the mill name, construction date and mill number usually are emblazoned on a chimney at every mill. These tall white structures with a black top are often the most prominent feature in the landscape. The mill buildings are beside the main road at Carlos Caraballo, as was No. 1550 when it was photographed here but one can only guess how often such a sight will be seen in the future.

Mill 440, Ifrain Alfonso – Standard Gauge

For big steam power on long trains, struggling up severe inclines, producing dramatic smoke effects, there can be only one line to consider in Cuba, and this is it. After taking *The Road to the Mills* off the Great Cuban Highway and driving through the pleasant little town of Ranchuelo, the first mill to be encountered is Ifrain Alfonso and what a fine array of majestic motive power is to be found here. The mill is home to five locomotives, all of which are large but two of which are enormous, and are by far the most massive machines on the whole of the island (pages 37 and 38).

Working practices have varied over the years, sometimes to include a diesel and sometimes not – one never knows what to expect until one arrives. There are wyes at the mill end, at the junction with the F.C.C. about a mile out from the mill, and also at the terminus at Pozo. The line has been truncated beyond here. In spite of the proliferation of wyes, locomotives rarely turn on this system. In any case, the mill-end wye has been used as a storage point for unwanted wagons for the last few seasons and cannot be used for locomotive turning, so outward trains of empties feature the locomotive running tender first, while the returning loads are hauled by locomotives running boiler first as seen on this and the next four pages.

Above, Ifrain Alfonso's 4-6-0 has slowed to a walking pace at the top of the incline before the final drop down to the mill and is seen coming out from behind the row of distinctive trees which also feature on pages 37 and 38. No. 1636, photographed on April 7, 1993, is an Alco Cooke of 1920 with the Works No. 62413 and is the mill's oldest locomotive – by all of five years! Built with 180 psi boiler pressure, a tractive effort of 23,800 lb., 50 in. driving wheels and 18 x 24 in. cylinders, the "Tenwheeler" weighed 120,000 lb. when built and presents a most handsome appearance. Originally carrying the number 2, it was delivered when the mill was still known as Santa Maria and has spent its entire working life here.

It is not a very regular performer however, perhaps because it is felt that the 4-6-0 wheel arrangement might not be the most suitable for the arduous duties unavoidably imposed on the locomotive stud at Ifrain Alfonso Mill due to the steep gradients here. Even the big 2-8-2s have been known, on occasion, to 'take-a-run' at the final climb and fail to breast the summit at the first attempt. They then have had to set back further to have another go, usually succeeding this second time but with even better smoke effects than the first! Lineside photographers are in their element here and most days during the zafra, groups of happy-snappers will be found at this, one of Cuba's most popular and spectacular mills.

Mill 440, Ifrain Alfonso – Standard Gauge

Unlike its stable mate opposite, No. 1637 is a comparative newcomer to Ifrain Alfonso Mill, having been built by Alco Cooke in 1925 with Works No. 66432. The locomotive was constructed as a speculative venture and was purchased by Elisha Atkins & Company for use in Cuba. It first went to San Germán and operated there as No. 11 until the mill's name was changed to Urbano Noris and the locomotive acquired its present designation in the all-Cuba numbering system. Records do not show when it moved to Ifrain Alfonso but Urbano Noris is in the east of the island and mills in that area were among the first to be dieselized, so it seems probable that the call to go west might have been made then.

No. 1637 is the 'commonest' of Ifrain's locomotives, being but a humble 2-8-0, yet it is able to hold up its head high here as the mill's only representative of that ubiquitous wheel arrangement. There are more "Consolidations" by far than any other sort of locomotive in Cuba, but not at Ifrain Alfonso, where the five allocated to this mill are of four different wheel arrangements, the others being a 4-6-0 (page 34), a 2-6-2 (page 36) and two 2-8-2s (pages 37 and 38). Common this 2-8-0 may be but it makes up for it by generally being kept very smart and always giving a good account of itself out on the line.

One of the youngest Alcos in Cuba (the very youngest, also of 1925, is Marcelo Salado's No. 1688), No. 1637 was built with 50 in. driving wheels, 18 x 24 in. cylinders and weighed in at 126,000 lb. when new, so is eminently suitable for coping with the demanding work at Ifrain Alfonso. Photographed running southwards through the hamlet of Horqueta with a train of loaded wagons on March 22, 1995, the locomotive's journey back to the mill is neither easy nor straightforward. The first obstacle to be encountered is the level crossing over the Great Cuban Highway (page 36), although as at Australia Mill (page 12 and 13), the crossing-keeper should stop oncoming road traffic but on more than one occasion has failed abysmally in this task and collisions have been known to occur.

There will be more loaded wagons to collect at the only intermediate loading point on the system and then the train will travel to a point just before the junction with the F.C.C, some three miles from the mill where it will stop for the driver to request permission to travel on F.C.C track. While he is doing this, water will probably be taken from the adjacent water tank, and then anything from an immediate restart to a wait of several hours might take place, depending on traffic on the mainline. MINAZ track will be regained via the wye at the bottom of the final climb and then the serious 'thrash' past the trees up to the summit will take place, with No. 1637 really showing what it is made of, often for the benefit of a battery of cameras.

Mill 440, Ifrain Alfonso – Standard Gauge

When the Russians abandoned Cuba in the early '90s, they left various projects uncompleted including a nuclear power-station, the Great Cuban Highway and a host of fly-overs and over-bridges, many of which were started and almost none of which were completed. One such bridge is shown in the picture on this page and is unusual in that the builders did actually get as far as connecting up one end to the road it was supposed to serve but not the other! This means that it is possible for photographers to get onto it to record the passage of trains such as this one, shown storming across the Great Cuban Highway on the level. From the top of the bridge, the sun will be 'right' for afternoon trains, and therefore 'wrong' in the mornings, which is one reason why No. 1635 was photographed from ground level on this occasion.

Road regulations in Cuba require drivers always to stop at level-crossings and this is why such locations invariably have street sellers in attendance; tourists' cars, which are distinctively marked, being the main targets of the opportunist tradesmen (rarely women!). The main product they try to unload on unsuspecting travellers is almost invariably garlic, so it ought not to come as much of a surprise to them that they do not sell much of it! But although they are no trouble to motorists, they are always there and in this picture a few of them can be seen on the right, keeping out of the way of the fast moving train.

The date is March 21, 1995 and the locomotive is one of very few 2-6-2s that work out on Cuba's 'main' sugar lines. Built by Baldwin in 1925 as the mill's No. 1, the locomotive carried the Works No. 58556, and had 18 x 24 in. cylinders with 44 in. driving wheels when built. This was before the mill was re-named Efrain Alfonso, a name it kept until the early '90s when another odd thing happened. It was decided that a partial renaming was appropriate, because at that time, observant visitors started noticing that the first initial 'E' was being changed to 'I' and it was very obvious where this had been done. The usual method employed was to paint a square over the original letter with bright fresh paint and then put the 'I' on top of that and so it has remained to this day, looking a real 'bodge-job' too!

Some spots have lettering which either cannot be reached without some difficulty as at the top of the mill-chimney or cannot easily be altered where there is a metal cut out name board at the main entrance. So, although the mill would appear to have undergone a name change, the old one is still visible to remind visitors what it was. Perhaps one day we shall discover why, after decades, it was thought necessary to go to all the trouble just to change an 'E' to an 'I' but this is Cuba and the chances are that we shall never know the answer to this and so many other questions in this wonderfully enigmatic country.

Mill 440, Ifrain Alfonso – Standard Gauge

One of the largest and youngest locomotives at work on any Cuban sugar line today is the 1935-built Baldwin No. 1850 (Works No. 61888), shown climbing past Ifrain Alfonso's distinctive collection of fine old trees just before the summit of the final climb up to the mill. The scene is lit by low evening sunshine on March 10, 1994 and one thing which is always pleasant in Cuba is the atmosphere which is virtually pollution-free. The low traffic levels on the roads are mainly responsible for this and although the sugar mills and the steam locomotives pour black smoke into the atmosphere through each zafra, it does not seem to affect the quality of the light at all.

Sunsets on the island are magical, both inland and on the coast, while even at midday when the sun is at its highest, reflected light still seems to reach parts of locomotives that the sun does not want to reach at this time of day in many other countries. It has been said that the light in Cuba is so good that even if a camera goes off by accident, the photographer will get a masterpiece and although this is obviously nonsense, there is no doubt that there is more than a grain of truth in it, or it would not have been said in the first place.

Despite being by some way the youngest Baldwin in the country, No. 1850 has nevertheless had a more varied past than most. It was constructed with 42 in. driving wheels and 19 x 26 in. cylinders yielding a tractive effort of 29,500 lb. for the Banes Railroad as their No. 112. It worked at Boston Mill, this later changing its name to Nicaragua. A move to Cristino Naranjo Mill followed and by the beginning of 1989 the locomotive had been set aside, its working life seemingly over. In March 1991, however, it came to Ifrain Alfonso where a long and difficult overhaul alongside stablemate No. 1910 (page 38) was commenced and by the 1994 season it was back at work hauling heavy trains along the undulating track.

As already mentioned, the former name here was Santa Maria and although all trace of this has vanished at the mill, the name is perpetuated at the nearby mainline station, located beside the wye where MINAZ tracks and the F.C.C. diverge. In normal circumstances therefore, sugar trains would not pass through this station, having to turn off to the right and climb the incline towards the mill. In 1997, ripe cane had to be cut at the far end of the system, even though Ifrain Alfonso Mill was not itself working. Its locomotives were, however, fully employed and, as the mill was out of action, visitors were treated to the rare sight of steam locomotives working through Santa Maria Station on their way to the nearest mill able to accept extra cane: Carlos Caraballo (page 33). Unfortunately, this special arrangement is not expected to last very long but it just goes to show that one never knows quite what to expect in Cuba.

Mill 440, Ifrain Alfonso – Standard Gauge

This is the largest working steam locomotive in Cuba and not only the one with the highest number but also the one whose return to service had been most eagerly awaited during the last few seasons. Arriving at Ifrain Alfonso in need of refurbishment at the same time as their other 2-8-2 No. 1850 (page 37), No. 1910 had been stored at the back of the shed in what looked like 'almost-ready' condition since the early 1990s. Each year, until 1996, it failed to appear but then railfans were delighted to discover that the last few jobs had been completed during the 'closed' season and it was back at work at last.

In 1996, the practice was for the 2-8-2s to work the line on alternate days with one of the other locomotives acting as yard pilot (i.e. three locomotives in steam) but in 1995 a diesel was in daily use with only one steam locomotive in operation. Prior to that, it was just as different each year but this is Cuba and, as often has to be pointed out, one never knows what to expect until one arrives at any particular mill. Just because it was working yesterday is no guarantee that it will be working today and just because a train of empties left yesterday at, say, 10:00 a.m., there is no guarantee that a train will leave at the same time today. Cuban sugar railways work just 'as required'.

No. 1910 is an Alco from the Brooks Works, Dunkirk, New York (Works No. 66284) and left there in January 1925 to work on the Cuba Railroad. It ran as No. 351 while on that line, having been built with Alco's usual 50 in. driving wheels, with a boiler pressure of 190 psi and with what are thought to be the biggest cylinders on any of the locomotives to work in Cuba – 22 x 28 in. These gave a tractive effort of 43,800 lb. At the same time as No. 351's delivery, two similar locomotives went to López Peña Mill (previously known as Baguanos) and one survives to be on show there. It is interesting to note the detail differences between the two giants, Nos. 1910 and 50.

One noticeable difference during the 1996 season was No. 1910's chimney-stack, not looking at all happy and decidedly not as it was built! The story goes that soon after reappearing and on an unrecorded date, No. 1910 was at one of Ifrain Alfonso's loading points when something came adrift from the machinery just as the locomotive was approaching and there was an unavoidable 'coming-together'! No doubt this was accompanied by much mirth as it would take a lot more than a bent stack to upset the average Cuban mill worker. Careful inspection of the photograph reveals that repairs had not been completed by the day it was taken: April 3. It remains to be seen ultimately what will be done but leaving it crooked may very well be the adopted option, permanently to remind everybody of what they should not be doing to Cuba's grandest.

Mill 443, Pepito Tey – 2'6" gauge

This is probably the most visited mill in all Cuba but not necessarily by railfans! Situated about eight miles due east of Cienfuegos (as the crow flies, but about double that distance by road), Pepito Tey Mill has been put on the 'Tourist Trail' by Cubatur, the state tourist organization, and it has been known for unsuspecting railfans to turn up at this quite delightful little centre, only to find it absolutely swarming with German, Italian or Canadian tourists!

There are beach package holiday resorts in the area and the large port of Cienfuegos itself attracts many foreign visitors, so what could be more natural than that they should be taken round a sugar mill? Pepito Tey is on the way from Cienfuegos to the show place town of Trinidad and besides being on UNESCO's world heritage list, Trinidad (founded in the 16th century) is also on most visitors' 'must see' list, making the narrow gauge steam operated sugar mill a convenient stopping off point en route.

There are 2'6" gauge lines at both ends of the mill, one winding its way round to a loading point near the main road about four miles away and the other quickly dividing into two much shorter lines with loading points at the end of each. The system is far from being the busiest in Cuba, but railfans love it for its smart little locomotives, typical of which is No. 1358, photographed at rest outside the shed on March 27, 1995. All are Baldwins and two or three are steamed on most days. The oldest dates from 1909 and the newest from 1924 but No. 1358 was delivered in 1915 with this manufacturer's 'standard' narrow gauge dimensions of 36 in. driving wheels, 14 x 20 in. cylinders and Works No. 42136. The overall design is nothing like standard however, as can be seen, and the elegantly tall stack, with spark-arrester, gives the impression of even earlier vintage.

Unusually, the running number of Pepito Tey's locomotives is not displayed on their cab sides, most of the allocation having the mill name here instead (CAI stands for Complejo Agro Industrial). Numbers are usually, but not always, on the smokebox door and in very small numerals on the back of the tender immediately under the light where it can just be made out. The tender sides are completely blank.

Pepito Tey Mill's name was Soledad when the little "Consolidation" came here new in 1915, running with the number 6 and the name *TERESA*. Yet another surprise is the presence of some 2-6-0s, probably not working, but 'on site' nevertheless. Locomotives have been 'abandoned' at various locations around the mill area, seemingly dumped but it is not wise to say that a locomotive has been dumped until it has all but disappeared, in view of what can be done to apparent wrecks. Suffice to say that there are a few locomotives 'awaiting repair' at Pepito Tey.

When No. 1358 was photographed, it was 'stone-cold'; just over an hour later it was at work switching wagons around the yard. When they want to, Cubans really can get things moving, and no mistake.

Mill 446, Carlos Baliño – Standard Gauge

Known as Ulacia Mill in pre-revolution days, Carlos Baliño has already been mentioned (page 30) because of the long roadside section which leaves it in a southerly direction and runs towards the little town of Santo Domingo. But it is not much good having great photographic possibilities if there are no trains to photograph and there has been a definite paucity of steam hauled trains along this line in the '90s. Some inter-mill workings have on rare occasions been noted between here and George Washington (not illustrated) but for much of the time, locomotives have generally been kept in steam for no obvious reason.

What yard work there is has been noted with the locomotive coupled smokebox to train and consequently visitors tend not to flock here in their thousands. Those few who do are likely to leave quickly on finding not much activity but this is a pity because the mill has a trio of fine 4-6-0s and the staff are very friendly and obliging. Two minutes before this photograph was taken on March 10, 1994, the locomotive was dead and deserted but the just arrived small group of railfans had obviously disturbed the crew's siesta and in no time at all there was smoke out of the stack and the crew were in the cab giving the impression of high activity. It was all a charade however and in another couple of minutes or so, No. 1555 and staff had resumed their slumbers. They were happy, the photographers were happy, everyone was happy, and look what an impressive sight this classic taper boilered "Tenwheeler" makes among all the paraphernalia outside its shed in the midday sunshine.

Carrying No. 1572 prior to 1987, No. 1555 is a Baldwin, built in 1918 with 56 in. driving wheels, 18 x 24 in. cylinders and the Works No. 50995. Its first home in Cuba was Cunagua Mill and the first number it carried was 11. But times change and these eventually became Bolivia and 1555. Busy Bolivia Mill (not illustrated) is now the home of four fireless locomotives, three Baldwins and an O&K which will usually reward a visit by being all in use simultaneously, and it may be that large locomotives like the 4-6-0 became redundant due to Bolivia's staff having a preference for fireless locomotives. It cannot, however, explain why No. 1555 was sent to Carlos Baliño, all of 150 miles away by rail!

The third 4-6-0 is No. 1643, recently arrived from George Washington. A fourth locomotive, a Vulcan Ironworks 2-8-0 of 1913, has been seen by visitors, lurking in the depths of the shed in a partly dismantled state and it may or may not be scheduled to work again. Unlike the other three residents at Carlos Baliño, it has been here all its working life. With so little work to do, it just might end up at a mill full of fireless locomotives if it is not careful!

Mill 448, Simón Bolivar, 2'3½" gauge

The second of the two mills 'joined' by the unusual gauge of 2'3½" (see pages 28 and 29), Simón Bolivar's own line heads off in a southeasterly direction from the mill, being part of the old Remedios-Chambas Railway as explained earlier. There is a junction in the village of Centeno controlled by a lone semaphore signal where a branch heads off south into the hills with road access by anything other than Jeep or tractor seeming to be almost impossible.

Photographed at the mill weighbridge on April 5, 1996, No. 1138 was a welcome sight for railfans as Simón Bolivar Mill had not been in use for the previous few seasons. There was only enough ripe cane in the area to keep a single mill in operation and Obdulio Morales was chosen to do all the work. Similar operating methods have been employed at two of Cuba's other 'adjacent' mills (pages 14 and 15 and page 20) where, since 1994, Granma has done all the processing and Victoria de Yaguajay has lain dormant although during some years the latter's locomotives have been in use and during some years they have not.

Simón Bolivar Mill is only four miles inland from the north coast of the island and the unusually gauged steam railways are not the only attractions for tourists in this part of Cuba. Nearby is a thriving colony of pink flamingos and, with literally hundreds of off shore islands both big and small, wildlife abounds in both the water and the air. But nowhere round here is very easy to reach so the fish and the birds remain relatively undisturbed.

No. 1138 is a 2-6-0 of 1908 vintage and has the Works No. 33067 but experts do not agree on its vital statistics, so they have been left unquoted. Despite being one of that rare breed, a narrow gauge "Mogul", its appearance is in general similar to the other 2'3½" gauge locomotives and generally dissimilar to most 2'6" gauge engines so its dimensions must remain in doubt. What is not in doubt is that it worked first for the North American Sugar Company at Narcisa Mill with the number 1 and the name *DICK*. Narcisa eventually became Obdulio Morales but, in common with many of the 2'3½" gauge locomotives, No. 1138 was interchanged with some from Simón Bolivar, the practise starting under the mill's original name of Vitoria and continuing through to the present day.

Both mills have their official allocation, and some locomotives have mill names on them but it is more usual to see *MINAZ* displayed as with their unique Cuban gauge, they can only work here at present! That does not mean that they can never work anywhere else, of course, as a little thing like being the wrong gauge is not likely to stop the Cubans from doing a quick regauging job if they put their minds to it but they would probably never need to as there is no shortage of spare narrow gauge locomotives in Cuba, just waiting to be employed.

Mill 503, Orlando González Ramirez – Standard Gauge

There are certain centres in Cuba where railfans feel they must stay in order for the 'best' mills to be within easy reach and Ciego de Avila is right up there at the top of the list with the other contender for "Best Stay" title being Santa Clara. It is quite difficult to decide which is the better of the two. For quantity, there are undeniably more steam mills in the Santa Clara area but for sheer quality Ciego de Avila would surely have to be declared the winner even if only on points. North of the town is Ciro Redondo Mill (front cover and page 48), southeast is Ecuador (pages 44 and 45), south is Venezuela (page 49) and due east is Orlando González Ramirez which is often abbreviated to O.G.R.

All four mills have lengthy systems with large locomotives hauling heavy trains over long roadside sections. All the mills seem set to retain steam traction for the foreseeable future. Staff are very friendly throughout the area, although visitors may not be allowed in some of the sheds unless they leave their cameras outside. One such is O.G.R. but why these restrictions should be imposed sometimes and not others has never been properly understood.

In this photograph, the building behind the Russian tractor on the right is the Traffic Office and this is where railfans head to find out when and where trains are running. All the information can be freely obtained and since the office overlooks the shed area, it is a convenient location for photographing the locomotives. This has never presented any problems but ask to enter the shed for the same purpose and permission is invariably withheld, albeit with a smile, but withheld nevertheless. Official visits to all Cuba's mill locomotive sheds can sometimes be arranged in advance through the Sugar Ministry in Havana but this itself can be a cumbersome and time consuming business so it is rarely thought worth the bother. Far better just to turn up, seek local permission while hoping for the best and being philosophical. If entry is allowed, that is good; if it is not, well, that is not the end of the world and there should be plenty of opportunities for good shots to be obtained away from the shed and out in the fields.

Here at O.G.R. is a case in point and the picture on this page is a good example. This is a loaded train being backed into the mill by the locomotive which collected it from the fields. No separate switchers here! Only the multitude of overhead wires detract at all from what would otherwise be a near perfect photographic location but this same problem is common on many railways throughout the world and photographers know it is something they will just have to live with. Here at the yard end of O.G.R. Mill there are more wires than at almost any other spot in Cuba and omitting them is impossible so photographers do not try. Better to let

them be part of the composition if possible than to forgo the picture purely on the grounds of obtrusive wires and poles.

The locomotive shown backing into the mill on March 26, 1995 is Vulcan Ironworks 2-8-0 No. 1836 of 1922 and has a very chequered history. Ordered by España Mill to be their No. 16, it never got there, going instead as No. 2 to Vertientes Mill and staying through the renaming to Panama. A move to Primero de Enero, one of the mills in the Ciego de Avila area that no longer uses steam locomotives, came next and the penultimate transfer followed when No. 1836 went up to the north coast to help out in the mill at Máximo Gomez.

Now the sole representative of the Vulcan Ironworks amongst Alcos and Baldwins at O.G.R., this 2-8-0, Works No. 3148, was built with 48 in. diameter driving wheels, 18 x 24 in. cylinders and a weight of 130,000 lb. A stalwart here for the past few seasons, this fine looking machine can usually be relied upon to perform well up and down the many lines and branches which constitute one of Cuba's more complex systems.

A short line runs northeast from the mill towards Guayacanes but most lines are generally to the south, one even eventually heading off east and connecting up with Venezuela's lines (page 49) more than 25 miles away. But the 'Jewel in the Crown' at O.G.R. is the street running through the pleasant and open little village of Limones Palmero seen on this page. This is the furthest point from the mill and at the far end of O.G.R's longest branch. A diesel has been seen working along here during 1996 and 1997 but so has steam and it is to be hoped that one of the last regular examples of true street running with steam power in Cuba will not succumb permanently to modern traction.

At Limones Palmero on March 26, 1995, was the 1920-built Schenectady Alco small-wheeled (50 in.) 4-6-0 No. 1736, Works No. 62655, exerting its 18 x 24 in. cylinders, 180 psi boiler pressure and 23,800 lb. tractive effort at the start of the 20 plus mile trip back to the mill with a long train of loads. There are no wyes in use on this system so attractive pictures can only be obtained during the return journey when locomotives run boiler first towards the mill and it was particularly rewarding to be able to include members of a Cuban family admiring the sedate passage of the "Tenwheeler" past their front door. With one previous owner, No. 1736 was No. 121 on the Cuban Central Railroad prior to arriving here and being required to perform tasks which are much more mundane than those it would have performed before. Photographers 'pacing' the trains on the long straight roadside stretches at the approach to the mill will not mind about this though, if conditions are right they will be more than happy that this good-looking 4-6-0 gives them the opportunity to exercise their shutters to such advantage.

Mill 504, Ecuador – Standard Gauge

Most unusually for Cuba, Ecuador Mill is not located in or near a town with the same name and, of the 31 mills illustrated, it is one of only three where the town and mill have different names. Rafael Freyre Mill is in Santa Lucia (Back Cover) and Boris Luis Santa Coloma (pages 10 and 11) is at Madruga.

Ecuador is located at Baragua which looks like any other little mill town although Ecuador Mill is different in that it produces refined as opposed to raw sugar. There are about a dozen refineries spread throughout the country and Ecuador is one of the last with a steam operated railway. At the raw sugar mills, the product will be seen being shipped out in large hoppers and it makes a change to see something different from cane wagons being switched by steam locomotives in readiness for despatch to the appropriate refinery, usually behind a large diesel.

While there are no timetables on Cuba's sugar lines, a few mills have operated what might be called a regular service and Ecuador is one such mill. Jose Smith Comas Mill (pages 18 and 19) has had a regular departure of empties round about 10:00 a.m. daily. Granma's locomotive has set off, light engine, from Carlos Rojas Station tender first in the region of 2:00 p.m. almost every day to collect loaded wagons from the far loading point (page 14) and at Ecuador, come dusk a succession of trains normally arrives at the yard to be switched into the mill by a waiting steam locomotive – possibly 2-6-0 Baldwin No. 1649 of 1919 which is seen in this shot. As the sun has lowered, so the activity has increased with trains coming in from both directions, although the northern line saw some diesel use during 1996 and 1997 but whatever the motive power was out on the line, the switching in the yard has normally been in the hands of steam locomotives.

With its original number 9, Agents Elisha Atkins & Company imported No. 1649 (Works No. 52423) into Cuba for the Punta Alegre Sugar Company on the north coast, where it continued under the mill's new name of Máximo Gomez until being redeployed at Ecuador, probably round about the time that No. 1836 left for service at Orlando González Ramirez Mill (page 42). Photographed in the yard on March 24, 1995, it has similar dimensions to the very smart No. 1550 at Carlos Caraballo Mill (page 33). These are 18 x 24 in. cylinders and 52 in. driving wheels but it is slightly more powerful because of its higher boiler pressure and has a tractive effort of 22,879 lb. It is one of two "Moguls" stabled here. Being the smallest of Ecuador Mill's locomotives, one or other has usually been deployed as the yard switcher, the larger locomotives being preferred for line work.

No wyes are in use at Ecuador so empties are delivered tender first which gives first time visitors the chance to learn the line and decide on photographic locations for when the trains are returning with the locomotive running with the smokebox leading. A shortage of bridges

and general lack of height can be very frustrating for photographers seeking new angles, so with thoughts heading ever upwards, loading points have come to be considered and a good time to seek permission to climb up onto one of these is during the outward tender first run. The photograph on page 29 was taken from such a spot and was the highest point on site being directly above the conveyor belt dropping cut lengths of cane into the railway car below.

The photographer was wobbling and shaking beside the completely unguarded mechanism and that photographs taken from such precarious positions come out as sharp as they do must say something for modern equipment, emulsions and techniques. Like all the illustrations in this book, it was taken with a Canon F1 35 mm camera on Ilford FP4 (Plus) film, a tried and trusted combination that has proved its worth over many years. Larger format cameras may have become all the rage during the '90s, but old habits die hard, and it has always seemed best to follow the old maxim: "If it ain't broke, don't fix it"! If it were obvious that results from larger cameras gave significant improvements over existing ones, then perhaps temptation might succeed, but as things are, 35 mm rules.

Where trains run past houses, householders can sometimes be persuaded to allow photographers to point their cameras out of upstairs windows, but in Cuba, most houses are single storey, with a covered patio like the one shown on the right hand side of the above picture, so going upstairs is rarely an option. There is no reason why ground-level shots should not be satisfying and the two lads standing outside their bungalow admiring the train give it a little bit more added interest. What the lads probably do not know is that the locomotive shown is now Ecuador Mill's biggest and one of Cuba's wonders. No. 1904 is a mighty "Consolidation" and was built by the Vulcan Ironworks of Wilkes-Barre in 1920, (Works No. 3102) being one of a pair of similar locomotives with 48 in. driving wheels, 19 x 26 in. cylinders and weighing 148,000 lb to be based here. Like its cousin No. 1835 (Alco, 1921), No. 1904's twin No. 1903 became disused by the early '90s. Nothing of use is wasted in Cuba, so when No. 1904 required attention what was more natural than that the serviceable boiler and cab from the Alco should be made to fit the frames of the Vulcan and that other odds and ends should be salvaged from No. 1903? The result is the fine looking hybrid 2-8-0 shown heading into the sunset on March 25, 1995 and it can truly be described as unique.

Starting out in life with the number 17 at Fructuoso Rodríguez (page 17) when that mill was called Limones, it moved to Violeta Mill (later to be called Primero de Enero but not illustrated) before coming to Ecuador. It would be a probable winner in the 'Most Rebuilt Locomotive in Cuba' competition. It is also a great tribute to the engineers and fitters who worked these wonders with little more than their bare hands and the will to succeed.

Mill 505, Carlos Manuel de Céspedes – Standard Gauge

Just off the main road that continues eastwards from where the Great Cuban Highway peters out on its way towards Holguín, Carlos Manuel de Céspedes Mill is no more than five miles west of the town of Florida (pronounced Flo-**ree**-der). This is a convenient stopping-off point for drivers heading towards the west end of the island. Unfortunately, steam operations here are in very serious decline and line work would appear to be exclusively in the hands of diesels but one or two locomotives have been steamed on a daily basis up to the present and it can only be hoped that things will get better and not worse as the collection of steam locomotives here really is something special.

Already mentioned is the little Davenport switcher recently brought out into the open, but four other manufacturers are represented at Céspedes and one of them is unique in Cuba so far as present knowledge extends. The 1896-built 4-6-0 No. 1567 is a product of the Pittsburgh Locomotive and Car Works, U.S.A., coming to this mill after having spent its early days on the Pittsburgh & Lake Erie Railroad. It is another of Céspedes' 'secret' locomotives in that it has not worked recently and is stuck in the shed, a shed into which visitors had not been encouraged to go until 1997!

This policy made it difficult to enjoy any of the rare machines here, another of which is a Porter "Mogul" of 1919 which once carried the number 6, but which since 1959 has been No. 1444. A third is No. 1445, a Baldwin of 1920 with the 2-6-2 wheel arrangement, never a very popular one in Cuba compared with the 2-6-0s and 2-8-0s which make up such a high proportion of the country's locomotives. A Vulcan Ironworks "Mogul" that was built as a 0-6-0 completes the non-Baldwin locomotives, but this maker is responsible for yet another rare sight which has seen some regular work in the '90s, No. 1173 (Works No. 42671), the 0-4-0ST+PT of 1915 shown on this page as it was on March 30, 1993.

A contender for the title 'The Most Disgusting Locomotive in Cuba', it has 33 in. wheels, tiny cylinders measuring only 10 x 16 in. and is completely dwarfed by the cane wagons it is asked to propel into the mill. The addition of oil-carrying panniers to the sides of the water carrying saddle tank gives No. 1173 a very odd appearance indeed, as can be seen.

Carlos Manuel de Céspedes Mill started life being named just Céspedes, and No. 1173 may well have been the first locomotive here as its old number was 1 but by now it should be quite used to its surroundings as it has never worked anywhere else which, when one comes to think of it, is not really surprising; who on earth would want to take delivery of such a filthy little switcher?!

The larger locomotives are in no better condition either, those not already described being Baldwins, as stated earlier. Shown here on the same date as No. 1173 is No. 1907, a

2-8-0 of 1924, built with 48 in. driving wheels, 19 x 26 in. cylinders, and a 180 psi boiler pressure providing a tractive effort of 29,918 lbs. The locomotive worked originally at Florida Mill (later to be named Argentina), running with the number 6, before making the move to Céspedes along with that mill's No. 7, now No. 1843.

Nos. 1907 and 1843 have consecutive Works Nos. 57865 and 57866 as they were ordered from Baldwin at the same time from the same batch. They look very similar so it is not unreasonable to suspect that they are the same, yet one is designated 19xx and the other 18xx! One wonders why? But no one in Cuba can, or will say why, and this is not the only example of strange numbering that has come to light under the all-Cuba numbering scheme.

On page 16, Cuba Libre's medium-sized 2-6-0 No. 1808 looks fine-and-dandy as it heads across an unguarded level crossing and out into the fields, but does it not look a little petite to be an '18'? A similar-looking locomotive is shown at Boris Luis Santa Coloma Mill on page 11 (No. 1604) and to all intents and purposes is identical. This pair differ by two power classifications as did the old and new numbers on the 2-6-2ST at Marcelo Salado Mill but these were eventually put right when No. 1160 was changed to No. 1343 in the late '70s or early '80s. It is probably much too late now to change a number merely to correct an anomaly, one suspects, and offending locomotives are more likely to be scrapped if they become troublesome. The spare parts would be invaluable and as there is no shortage of replacements up and down the country, mill mechanics and engineers would be given the chance to hone up their skills by proving that even rusty wrecks can be brought back to life, usually against all the odds.

The three other incumbents at Carlos Manuel de Céspedes Mill are 2-8-0s and are late (for Cuba) examples from 1919 and 1920, each of a different power classification and with a different history. No. 1746 was No. 16 at Pilar Mill, No. 1838 is an ex mainline locomotive, coming from the F.C. Norte de Cuba with the number 48 and No. 1906 is a lifelong Céspedes locomotive which first carried the number 5.

Attempting to trace the histories and technical details of all the locomotives of Cuba has been an immense project in itself and we must be eternally grateful to the dedicated few on both sides of the Atlantic who have devoted so much of their time and effort over the years to this task, often unearthing obscure details relating to locomotives which are even more obscure. Obviously this has been very much a labour of love and has been done more for personal satisfaction than for financial reward but it is to be hoped that all this information will have added greatly to the value of the seemingly timeless display of illustrations that fill the bulk of this publication.

Mill 515, Ciro Redondo – Standard Gauge

For big locomotives in quantity, Ciro Redondo Mill is the best place to be in Cuba. Drivers of 50 in. diameter and 20 x 24 in. cylinders are the 'norm' at Ciro Redondo, the mill with the big, big locomotives.

Renowned throughout the country and much of the rest of the world for its magnificent set of matching "Consolidations", their number is being reduced slowly from the nine it had at the turn of the decade by the need to cannibalize spare locomotives to keep the rest of the fleet going. By 1997, the number left in working order was down to about a half-dozen but hope springs eternal and the ever-resourceful Cubans may find a way of keeping this number in use without having to resort to the breaker's torch.

Odd one out Baldwin No. 1651 has been out of use for some years now, but the rest not only look virtually the same, they also have consecutive running numbers from 1826 to 1834. This is despite the facts that No. 1826 is an Alco Cooke, that all the rest are Baldwins and that their histories are not all the same! Five came via the Cuba Cane Sugar Company to Ciro Redondo, three were first on the books of the Cuban Trading Company and one, No. 1833, was at the now defunct Santo Tomas Mill before being transferred to Ciro Redondo.

That Ciro Redondo's 2-8-0s might collectively be regarded as the most impressive in Cuba is reflected by the choice of photograph for the Front Cover. Seen on March 23, 1995, No. 1832 (Works No. 53952) presents a sight as typical of Cuba's sugar railways as is possible to obtain. The volumes of thick black smoke are normally there wherever there are steam locomotives and, if not, they soon are if photographers request them! Also typical are the overgrown track and the hard-working locomotive; a dirt track running along beside it through endless fields of sugar cane; workers standing all over the tender – no combination could better exemplify steam operations on the 'Paradise Island'.

The photograph above shows 'twin' No. 1828 (Works No. 52970) easing a lengthy loaded train homewards across the F.C.C. mainline at the flat-crossing south of the mill on March 15, 1994. On the left is the crossing signal cabin, part of a 'private' residence occupied by a friendly Cuban family, most of whom have come out into their front yard to savour the passage of the mighty "Consolidation" and its train. This has been held here for a few minutes whilst 'orders' have been sought, and received. With the knowledge that the mainline is now free, No. 1828 is on its way to disgorge its precious cargo. So impressive is the whole experience that even a lady cyclist on the right has stopped and turned round so as not to miss any of the action! The switchman on the front of the locomotive will soon be jumping down to change the switch at the yard so that the locomotive can then be separated from the train and head off towards the shed for a well deserved rest.